Stochastic Evolutions of Dynamic Traffic Flow

Xiqun (Michael) Chen
Li Li · Qixin Shi

Stochastic Evolutions
of Dynamic Traffic Flow

Modeling and Applications

Xiqun (Michael) Chen
University of Maryland
College Park, MD
USA

Li Li
Qixin Shi
Tsinghua University
Beijing
China

ISBN 978-3-662-44571-6   ISBN 978-3-662-44572-3   (eBook)
DOI 10.1007/978-3-662-44572-3

Jointly published with Tsinghua University Press, Beijing
ISBN: 978-7-302-37479-4 Tsinghua University Press, Beijing

Library of Congress Control Number: 2014948776

Springer Heidelberg New York Dordrecht London

Printed on acid-free paper

Springer is part of Springer Science+Business Media (www.springer.com)

To Dandan, Cynthia, Yuexin and Shuhuai
The families of Xiqun (Michael) Chen

Preface

Road traffic flow is intrinsic with stochastic and dynamic characteristics so that traditional deterministic theory no longer satisfies requirements of the evolution analysis. Stochastic traffic flow modeling aims to study relationships of transportation components. The kernel is an investigation of both stochastic characteristics and traffic congestion evolution mechanism using headway, spacing, and velocity distributions. The primary contents include empirical observations, connections with microscopic and macroscopic traffic flow models, and traffic breakdown analysis of highway bottlenecks.

The book first analyzes characteristics of empirical traffic flow measurements to reveal the underlying mechanism of complexity and stochastic evolutions. By using *Eulerian* measurements (e.g., inductive loop data) and *Lagrangian* measurements (e.g., vehicular trajectory data), we study headway-spacing-velocity distributions quantitatively and qualitatively. Meanwhile, disturbances of congested platoons (jam queues) and time-frequency properties of oscillations, which establish the empirical foundation for stochastic traffic flow modeling.

Then we establish a Markov car-following model by incorporating the connection between headway-spacing-velocity distributions and microscopic car-following models using the transition probability matrix to describe random choices of headways/spacings by drivers. Results show that the stochastic model more veritably reflects the dynamic evolution characteristics of traffic flow. As discussions of the connection between headway-spacing-velocity distributions and the macroscopic fundamental diagram model, we analyze the probability densities and probabilistic boundaries of congested flow in flow-density plot by proposing a stochastic extension of Newell's simplified model to study wide scattering features of flow-density points.

For applications to highway on-ramp bottlenecks, a traffic flow breakdown probability model is proposed based on headway/spacing distributions. We reveal the mechanism of transitions from disturbances to traffic congestion, and the phase diagram analysis based on a spatial-temporal queueing model that is beneficial to obtain optimal control strategies to improve the reliability of road traffic flow.

We would like to acknowledge the following people for their contributions in bringing this book to completion. We are grateful to Prof. Meng Li and Prof. Zhiheng Li for their sustaining guidance and encouragement. They share perspectives on dynamic transportation planning and traffic control approaches for the analysis of stochastic traffic phenomena. It has been a privilege for us to work with them. Thanks to Profs. Huapu Lu, Jing Shi, Xinmiao Yang, and Ruimin Li, who provided a number of valuable comments and suggestions that substantially improved this book.

We would like to show our gratitude to Prof. Wei-Bin Zhang, Drs. Liping Zhang, Kun Zhou, and Jing-Quan Li for their kind help in the *FHWA—Advanced Traffic Signal Control Algorithms* project. They studied the dynamic all-read extension and minimized the yellow arrival rate using vehicular trajectories, connected vehicles and eco-driving challenges on urban arterials while at California PATH, University of California at Berkeley. In addition, the book was inspired by the work of Prof. Pravin Varaiya on the exploration of magnetic sensors for travel time analysis, and Prof. Alex Skabardonis's work on PeMS for traditional loop data applications. The book has been advanced by several discussions with Prof. H. Michael Zhang, who pointed toward the deeper understanding of stochastic fundamental diagram and traffic breakdown probability model.

We also thank our former students and friends both in research and in life at Tsinghua University; they are Jin Wang, Xuyan Qin, Zhen Qian, Yunfei Sha, Liguang Sun, Runhua Qian, Jinxin Cao, Chun Xu, Shen Dong, Xinxin Yu, Xiongfei Zhang, Yuan Fang, Xiangqian Chen, Jia Wang, Chenyi Chen, Weijun Xie, Zhaomiao Guo, Yan Yan, Lijun Sun, and all other students in our program. You have provided us support on a number of research projects, brainstorming ideas in seminars or collaboration on articles. We also sincerely appreciate the friendship and generous help from my colleagues in Institute of Transportation Studies, University of California at Berkeley, Weihua Gu, Bo Zou, Yiguang Xuan, Haotian Liu, Xiaofei Hu, Lu Hao, Haoyu Chen, and Yang Yang.

Finally, Dr. Xiqun (Michael) Chen sends special thanks to his parents Yuexin Chen and Shuhuai Guo, his wife Dandan Li, his daughter Cynthia Chen, who will be born in the autumn of 2014, and elder sister Xili Chen. It would not have been possible for him to finish this book without their love and unwavering support.

May 2014   Xiqun (Michael) Chen
                                                                                                        Li Li
                                                                                                   Qixin Shi

Contents

Acronyms

Mathematical Symbols

a, b	Scale and translation parameters of Wavelet Transform
$a_{\max}, v_{\max}$	Maximal acceleration and maximal speed
$a_n(t)$	Acceleration of the nth vehicle at time t
A	Perturbation matrix
$\{\mathcal{A}_i, \ i \in \mathbb{N}^+\}$	Mutually exclusive events
b_i	Deceleration of the following vehicle
c_0	Substitution of speed
C_h	Coefficient of variation
$\mathcal{C}, \mathcal{C}', \hat{\mathcal{C}}$	Copula, its density function and empirical copula
d_i	Spacing of full stop
$E(\cdot)$	Average wavelet energy
$\mathrm{E}[\cdot], \ \mathrm{Var}[\cdot], \ \sqrt{S^2[\cdot]}$	Expectation, variance and standard deviation operators
$f(h\|\alpha, \beta, h_0)$	PDF of Gamma distribution
$f(h\|\lambda)$	PDF of negative exponential distribution
$f(h\|\lambda, h_0)$	PDF of shifted exponential distribution
$f(h\|\mu_h, \sigma_h, h_0)$	PDF of lognormal distribution
F, F_1, F_2	CDF
$F(h\|\alpha, \beta, h_0)$	CDF of Gamma distribution
$F(h\|\lambda)$	CDF of negative exponential distribution
$F(h\|\lambda, h_0)$	CDF of shifted exponential distribution
$F(h\|\mu_h, \sigma_h, h_0)$	CDF of lognormal distribution
$\hat{F}, \hat{F}_1, \hat{F}_2$	Empirical CDF
$F_c(q_S)$	CDF of upstream flow q_S, $1 - F_c(q_S)$ is lifetime function
$\widehat{F}_c(q_S)$	PLM estimation

$F_{H\|\upsilon}$, $F_{S\|\upsilon}$	Conditional distribution of headway/spacing with respect to υ
$F_{H,\alpha\|\upsilon}^{-1}$	α quantile of headway distribution
$\bar{h}$	Average headway
h_0, $\hat{h}_0$	Shift coefficient and its estimate
$h_{0,p}$, $h_{0,q}$	Shift coefficients of car-following state and free flow state
h_{free}	Critical headway of free flow
h_i	Observation samples of headway, $i = 1, 2, \ldots, n$
$h_k(t)$	Headway
$h_{\min}$	Minimal headway
h_{random}	Random headway from uniform distribution $\mathcal{U}(H_j^-, H_j^+)$
$\tilde{h}$, $\tilde{\upsilon}$	Mean headway and speed of $\tilde{\Upsilon}$
$\tilde{h}_{n,n-1}$, $\tilde{v}_n$	Mathematical expectations of headway and speed
$H(\cdot)$	Heaviside function
H_i^-, H_i^+	Upper and lower limits of the ith headway state
H_n	Summation of headway
$H_n^{(1)}$, $H_n^{(2)}$	Disturbance propagation time to the nth vehicle in ac/deceleration waves
$i = \sqrt{-1}$	Imaginary unit
$\mathcal{J}$	Population
$\mathcal{J}_{ij}$	Sample size of that $h_k(t)$ belongs to state i and $h_k(t + \Delta t)$ belongs to state j
$\mathcal{J}_k$	Number of triple, analogically, $\mathcal{J}_k^*$, $\mathcal{J}_k^-$, $\mathcal{J}_k^+$, $\mathcal{J}_k^\Delta$
k_1, k_2	Weight coefficients of density and speed gradients
L	Circular road length
L_{cong}	Jam queue length
L_{free}	Free flow traffic length
L_{qs}, $L_{\rho,v}$	Likelihood function
$\mathcal{L}$	Lagrangian function
$\mathbb{N}$, $\mathbb{M}$	Finite dimensional state space, satisfying $\mathbb{N} = \{1, 2, \ldots, n\}$, $\mathbb{M} = \{1, 2, \ldots, m\}$
$\mathbb{N}^+$	The set of positive integers
$\mathcal{N}$, Log$-\mathcal{N}$	Normal and lognormal distributions
$o(\cdot)$	Infinitesimal of higher order
$p(h)$	PDF of car-following state headway
$p^*(\lambda)$	Laplace transform of $p(h)$
p_{slow}	Slow-to-start probability
$\hat{p}_k$, $\bar{p}$	Probability estimate
P_B	Traffic flow breakdown probability
$\tilde{P}_0(v)$	Distribution of speed expectation

$\tilde{P}(v\|x,t)$	Speed distribution
P_1, P_2	Pressure terms, satisfying $\partial_\rho P_1 \leqslant 0$ and $\partial_v P_2 \leqslant 0$
P_{GUE}	Gaussian unitary ensemble, GUE
$P_{Poisson}$	Poisson distribution
$\boldsymbol{P}$	Transition probability matrix, satisfying $\boldsymbol{P} = (P_{ij})$
$\boldsymbol{P}_l$, π_l	Transition probability matrix and stationary distribution of headway in the ℓth velocity range
$\mathcal{P}(n,t)$	Probability of a jam queue with n vehicles at time t
$\mathcal{P}_{(z,t)}$	Probability of being state z at time t
$\mathcal{P}_{\mathcal{S}}(t)$	Probability of being state $\mathcal{S}$ at time t
$q(h)$	PDF of headway in free flow state
$q_e(\rho)$, $v_e(\rho)$	Equilibrium functions of flow and speed
q_{lower}, q_{upper}	Maximal and minimal traffic flow rates, analogically, ρ_{lower}, ρ_{upper}
q_{main}, q_{ramp}	Mainline flow and ramp flow
q_{max}	Maximal flow rate
$\bar{q}_S$	Breakdown traffic flow rate
r	Iteration times
R	Least squares residual
$\mathcal{R}$	Response function
$s^A(v)$, $s^D(v)$	Spacings of ac/deceleration curves at speed v
s_{before}, s_{after}	Spacings when joining and departing from jam queues
s_{break}, s_{free}	Spacing thresholds of braking and free flow states
s_{cong}	Spacing in a jam queue
s_{start}, s_{stop}	Spacing thresholds of starting and stopping states
$s_i(t)$	Spacing
s_{min}, s_{max}	Minimal and maximal critical spacings
$s_{n,n-1}$, $h_{n,n-1}$	Headway and spacing of the nth vehicle
s^A_{stop}, s^D_{stop}	Fully stopping spacings of ac/deceleration curves
s^A_{upper}, s^D_{upper}	Maximal and minimal spacings corresponding to the maximal speed $\tilde{v}^+$ in metastable traffic flow, analogically, s^A_{lower}, s^D_{lower}
$\mathbb{S}_k$	Sample set, analogically, $\mathbb{S}^*_k$, $\mathbb{S}^-_k$, $\mathbb{S}^+_k$, $\mathbb{S}^\Delta_k$
$\mathcal{S}$	Arbitrary discrete state of traffic flow
S^A, S^D	Probabilistic boundaries of microscopic fundamental diagram
t	Time
$\left(t^{(k)}_{a,i},\ x^{(k)}_{a,i}\right)$	Acceleration point of the ith vehicle, analogically, $\left(t^{(k)}_{d,i},\ x^{(k)}_{d,i}\right)$
T	Entering time interval of ramping vehicles
$\bar{T}$	Average travel time

$\mathcal{T}, \mathcal{T}^-, \mathcal{T}^+$	Tau factor
$\mathcal{T}^*, \bar{\mathcal{T}}^*$	Tau factor Related statistics
$\hat{\mathcal{T}}_k^-, \hat{\mathcal{T}}_k^+, \hat{\mathcal{T}}_k^\pm$	Estimate of Tau factor
v, u	Velocity
$v(x, t)$	Mean velocity at location x and time t
$v_{\text{cong}}, v_{\text{free}}$	Congestion wave speed and free flow speed
$v_n(t)$	Speed of the nth vehicle at time t
$v_{n,\text{safe}}$	Lower bound of safety speed
$v_{\text{opt}}(x_n(t))$	Optimal speed function
$v + c_\pm$	Characteristic speed
$\bar{v}(x, t)$	Expected speed
$\tilde{v}, \tilde{\rho}$	Perturbation amplitudes of speed and density
w	Congestion wave speed, analogically, w_i, $-w_i$, $-W_n$
$w(t - k)$	Window function
w^A, w^D	Ac/deceleration congestion wave speeds
$\mathcal{W}$	Weibull
x	Location
$x_n(t)$	Location of the nth vehicle at time t
$\tilde{X} = (\tilde{\rho}, \tilde{v})^T$	Column vector of perturbation amplitude
$(x, v) \mapsto (y, u)$	Transition rate from state (x, v) to state (y, u) of traffic flow
z, z'	Arbitrary continuous traffic flow states
(Z_1, Z_2)	Bivariate random variables
$(Z_1^{(i)}, Z_2^{(i)})$	Observations of bivariate random variables, $i \in \mathbb{N}$
$\{z_t\}$	Traffic flow discrete time series data, $t = 0, 1, \ldots, T - 1$
Z_n	Summation of random variables
$Z_{n,\text{typical}}$	Extreme points of PDF of Z_n
α, β	Parameters of Weibull distribution
$\hat{\alpha}, \hat{\beta}$	Parameter estimations of Weibull distribution
$\gamma_i^S, \gamma_i^B, \gamma_i^E$	Net spacings before, within and after the deceleration of the ith vehicle
Γ	Gamma function
$\delta v, \delta \rho$	Speed and density variations
Δ	Field data measurement interval
Δt	Update time step
$\Delta v_n, \Delta x_n$	Speed and location differences between the nth and the $n - 1$ the vehicles
$\epsilon(x, t)$	Noise function at location x and time t
$\eta(\rho, v)$	Inertial coefficient of driving behaviors
$\eta_{\text{lower}}, \eta_{\text{upper}}$	$\alpha/2$ and $(1 - \alpha/2)$ percentiles of standard normal distribution

θ_1, θ_2	Substitution parameters
$\theta_{\mathrm{OCT}}, \theta_{\mathrm{HCT}}$	Dimensionless critical coefficients
$\boldsymbol{\theta}$	Indicator function, analogically, $\mathbf{1}\,(\cdot) \mapsto \{0,\ 1\}$
$\boldsymbol{\Theta}(x,t)$	States and parameter vector at location x and time t
$\vartheta_i^-, \vartheta_i^{+}$	Upper and lower bounds of the ith vehicle headway
$\kappa, \tilde{\kappa}$	First vehicle delay after perturbations
$\lambda, \tilde{\lambda}$	Parameters and their estimations
λ_p, λ_q	Parameters of car-following and free flow states
μ, σ	Lognormal distribution parameters, analogically, headway $(\mu_h,\ \sigma_h)$, spacing $(\mu_s,\ \sigma_s)$, reaction time $(\mu_\tau,\ \sigma_\tau)$, bivariate lognormal distribution $(\mu_{z_1|z_2}, \sigma_{z_1|z_2})$
$\hat{\mu}, \hat{\sigma}$	Lognormal distribution parameter estimation
ξ, ζ	Random variable
$\xi(\rho, v)$	Anticipation coefficient of driving behavior
$\rho(x,t)$	Average density at location x and time t
ρ_0	Steady state density, initial density
ρ_z	Correlation coefficient
$\tilde{\rho}(x,t,v)$	Phase space density
$\zeta, \bar{\zeta}$	Vehicle gap and average vehicle gap
τ	Latency time, relaxation time and reaction time
$\tau_{\mathrm{in}}, \tau_{\mathrm{out}}$	Interarrival time and service time, analogically, $\tau_{\mathrm{in},i}$, $\tilde{\tau}_{\mathrm{in}}^{(k)}$, $\tilde{\tau}_{\mathrm{out}}$, $\hat{\tau}_{\mathrm{out}}$
$\tau_{\mathrm{in}}(m)$	Interarrival time summation of m vehicle
$\tau_i^{\mathrm{A}}, \tau_i^{\mathrm{D}}$	Reaction time of ac/deceleration
$(\tau^{\mathrm{A}},\ w^{\mathrm{A}}\ s_{\mathrm{stop}}^{\mathrm{A}})$	Characteristic parameters of acceleration curve, analogically, $(\tau^{\mathrm{D}},\ w^{\mathrm{D}}\ s_{\mathrm{stop}}^{\mathrm{D}})$
$\tilde{\Upsilon}, \Upsilon_k$	Inhomogeneous platoon and homogeneous sub-platoon
$\phi_1, \phi_2, \varphi_1, \varphi_2$	Substitution parameters
$\Phi(\cdot)$	Standard normal distribution
φ	Proportionality coefficient
$\varphi_{FF},\ \varphi_{PLC}$	Dimensionless critical coefficient
$\psi(t),\ \hat{\psi}(f)$	Mother wavelet function and its Fourier transform, the conjugate function is $\hat{\psi}^*(f)$
ω	Digital frequency
$\omega(k),\ k \in \mathbb{N}^+$	Angular frequency
$\omega_{zz'}$	Transition rate from state z to z'
$\omega_+(n),\ \omega_-(n)$	Transition rate of jam queue length from $(n \mapsto n+1)$ and $(n \mapsto n-1)$
$\omega_\pm$	Complex solution of ω
$\omega_{\mathcal{S}\mathcal{S}'}$	Transition rate from state $\mathcal{S}$ to $\mathcal{S}'$

Ω	Analog frequency
$\ell(\cdot)$	Likelihood function
$\Re, \Im$	Real part and imaginary part
$\emptyset$	Empty set

Abbreviations

ACTM	Asymmetric Cell Transmission Model
A-curve	Acceleration Curve
CA	Cellular Automaton
CCTM	Compositional Cell Transmission Model
CDF	Cumulative Distribution Function
CTM	Cell Transmission Model
CWT	Continuous Wavelet Transform
D-curve	Deceleration Curve
DFT	Discrete Fourier Transform
DTA	Dynamic Traffic Assignment
DWT	Discrete Wavelet Transform
EKF	Extended Kalman Filter
ELCTM	Enhanced Lagged CTM
EM	Error Mean
FD	Fundamental Diagram
FF	Free Flow
FHWA	Federal Highway Administration
FT	Fourier Transform
G/D/1	General Determinant 1
G/G/1	General General 1
GKT	Gas-Kinetic-based Traffic Model
GUE	Gaussian Unitary Ensemble
HCT	Homogeneous Congested Traffic
i.i.d.	Independent and Identically Distributed
IDM	Intelligent Driver Model
ITS	Intelligent Transportation System
K–S	Kolmogorov–Smirnov Test
LCTM	Lagged Cell Transmission Model
LPO	Log Periodic Oscillations
LSCTM	Location Specific CTM
LSR	Least Squares Regression
MCTM	Modified Cell Transmission Model
MLC	Moving Local Cluster
NGSIM	Next Generation Simulation
OCT	Oscillated Congested Traffic
OVM	Optimal Velocity Model

PA	Perturbation Analysis
PDF	Probability Density Function
PeMS	Performance Measurement System
PLC	Pinned Local Cluster
PLM	Product Limit Method
RMSE	Root-Mean-Square Error
RMT	Random Matrix Theory
SCTM	Stochastic Cell Transmission Model
SSM	State Switching Model
STFT	Short-Term Fourier Transform
TF-BP	Traffic Flow Breakdown Probability
TSG/SGW	Triggered Stop-and-Go Waves
WSS	Second-order Wide-sense Stationary Process
WT	Wavelet Transform

Chapter 1
Introduction

1.1 Motivation

Traffic congestion results in a number of negative effects on: (1) *Mobility*. Travel delays and wasting time of passengers or goods reduce the efficiency of transportation systems and increase opportunity costs; (2) *Safety*. Higher probability of serious injuries and death crashes as a result of human fallibility in congested flows; (3) *Sustainability*. Increased travel time and oscillatory acceleration/braking maneuvers in traffic congestion induce significant environmental impacts, such as fuel consumption, greenhouse gas emissions, air pollution, noises, etc.

Road traffic flow is influenced by various random factors, including both external factors (e.g., weather) and internal factors (e.g., transportation facilities, vehicle characteristics, driver behaviors, etc.). These stochastic influences make deterministic traffic flow models difficult to accurately estimate and predict dynamic evolutions. To overcome this problem, numerous stochastic approaches were developed for continuous traffic flow on the basis of microscopic/macroscopic traffic flow models. Particularly, different kinds of drivers (e.g., aggressive versus passive, young versus old, skilled versus greenhand, rigorous versus fatigued) run different kinds of vehicles (e.g., cars versus trucks, buses) on the same road, and thus, traffic flow is heterogeneous.

Since headway/spacing/velocity perform fundamental roles in stochastic traffic flow modeling, it is significant to study their stochastic characteristics in traffic flow evolutions. According to *Highway Capacity Manual 2000* (on page 48 of Transportation Research Board of the National Academies (2000)),

Definition 1.1 Headway (time headway, h) is the time, in seconds, between two successive vehicles as they pass a point on the roadway, measured from the same common feature of both vehicles.

Studies on headway distributions received continuous interests since the birth of traffic flow research, because of their wide applications ranging from measuring road capacity to scheduling traffic signals. Headway distribution model is one of

© Tsinghua University Press, Beijing and Springer-Verlag Berlin Heidelberg 2015
X. (M.) Chen et al., *Stochastic Evolutions of Dynamic Traffic Flow*,
DOI 10.1007/978-3-662-44572-3_1

the most well-known mesoscopic flow models, in which vehicles and driver behaviors are described by more aggregated terms instead of individual interaction rules. It is assumed that these headways are independent and identically distributed (i.i.d.) random variables.

Spacing is another important statistics in traffic flow modeling. According to *Highway Capacity Manual 2000* (on p. 56 of Transportation Research Board of the National Academies (2000)),

Definition 1.2 Spacing (space gap, s) is defined as the distance, in feet or meters, between two successive vehicles in a traffic lane, measured from the same common feature of the vehicles.

Numerous probability models have been proposed to fit the empirical distributions of headway/spacing (Branston 1976; Buckley 1968; Cowan 1975; Hoogendoorn and Bovy 1998; Kerner and Klenov 2006; Luttinen 1996; Ovuworie et al. 1980; Thiemann et al. 2008; Tolle 1976; Wasielewski 1974; Zhang et al. 2007), such as Gaussian distribution, lognormal distribution, etc. Based on the statistical characteristics, we incorporate a Markov model that yields the steady-state headway/spacing distributions as those observed in practice to simulate transient-state statistics of road traffic for different driving scenarios, including highway traffic and urban street traffic. This stochastic model is in accordance with our daily driving experiences.

Traffic flows are often described by three macroscopic variables: traffic flow rate q (veh/h), traffic density ρ (veh/km), and mean velocity v (km/h). The relationships among the macroscopic variables and the aforementioned microscopic variables are $q = 3600/h$ and $\rho = 1000/s$. The fundamental diagram (FD) is the most widely used model to describe the equilibrium flow-density relationship. But field observations show the possible inconformity between the wide scattering features of flow-density plot and the fundamental diagram assumption. Thus, it is necessary to establish a tight link and a statistical relationship between the spacing/headway distribution model and the wide scattering feature of flow-density plot because the stochastic fundamental diagram is more consistent in empirical measurements.

Traffic congestion performs temporal-spatial features observed from field traffic data. Many random factors influence empirical observations, such as driver characteristics, stochastic car-following behaviors, random headway/spacing perceptions and choices, weather and infrastructure conditions, time of day, etc. In particular, traffic congestion caused by on-ramp flow is one of the main influencing factors for highway mobility. From the macroscopic point of view, traffic breakdown phenomena show sudden transitions from free flow states to congested states (obvious velocity reduction) within a short period of time. The problem to describe the formation, evolution and dissipation of such congestions received increasing interests in the last decade. Since phase diagram approach provides a tool to categorize the rich congestion patterns and explain the origins of some complex phenomena at bottlenecks. To solve this problem via a minimum number of parameters and rules, we may need a concise temporal-spatial queueing model to depict traffic jams in trajectories by extending the Newell's simplified model.

1.2 Objectives

This book aims to investigate the stochastic evolution modeling for highway traffic flow to reveal the mechanism of complex and dynamic traffic congestion. We will first review the existing studies on stochastic traffic flow approaches and probabilistic distributions of headway/spacing. Motivated both by the availability of vehicle trajectory data, and by the need for traffic congestion reduction, the goal of this book is to develop a framework of stochastic traffic flow evolution analysis by using *Eulerian* and *Lagrangian* field measurements that empirically reveal the important characteristics of traffic congestion.

We aim to emphasize the features of headway/spacing/velocity distributions, which establish a tight connection between microscopic and macroscopic traffic flow approaches. We also target to adapt these empirical observations into stochastic modeling problems by proposing a Markov car-following model and a stochastic fundamental diagram model, because the randomness explicitly embedded in these models could be reasonably explained as the outcome of the unconscious and also inaccurate perceptions of space and/or time interval that people have. We will study a behavioral or psychological mechanism that is different from other approaches to explain the stochastic traffic flow features.

We aim to develop a traffic flow breakdown probability model and its corresponding phase diagram approach to explore the phenomena that occur with continuous oscillations and lead to a wide range of spatial-temporal traffic congestion near on-ramp bottlenecks. We will study the transition process from perturbations to traffic jams in the metastable traffic flow, and the varying features of traffic flow breakdown by using vehicle trajectory data. This model will benefit the optimal control models of active traffic management by minimizing traffic flow breakdown probability and maximizing the expectation of stochastic traffic capacity.

1.3 Contributions

This book contains several contributions to the state of art in stochastic traffic flow modeling based on headway/spacing distributions.

- **A microscopic unified Markov-process car-following model: stochastic behaviors in unconscious and inaccurate perceptions of headway/spacing.**
  We link two research directions of road traffic, the mesoscopic headway distribution model and the microscopic vehicle interaction model, together to account for the empirical headway/spacing distributions. A unified car-following model (a Markov-headway model for high velocity and a Markov-spacing model for low velocity) will be proposed to simulate different driving scenarios, including traffic on highways and at intersections. The parameters of this model are directly estimated from the Next Generation Simulation (NGSIM) trajectory data. Empirical headway/spacing distributions are viewed as the outcomes of stochastic

car-following behaviors and the reflections of the unconscious and inaccurate perceptions of space and/or time intervals that people may have. This explanation can be viewed as a natural extension of the well-known psychological car-following model (the action point model). The psychological explanation by the asymmetric stochastic extension of the Tau theory will be also presented. Furthermore, the fast simulation speed of this model will benefit transportation planning and surrogate testing of traffic signals.

- **A macroscopic stochastic fundamental diagram model: characterizing scattering features in flow-density plots using a stochastic platoon model.**
  Fundamental diagram is observed as wide scattering in the congested flow regime that requires a stochastic mechanism to explain this feature. Based on the Newell's simplified car-following model, we discuss the implicit but tight connection between the microscopic spacing/headway distributions and the macroscopic scattering feature of flow-density plot. We examine microscopic driving behaviors that are retrieved from the NGSIM trajectory database and study the asymmetric driving behaviors that result in a family of velocity-dependent lognormal type headway/spacing distributions. Then, we propose a stochastic platoon model to characterize the distribution of points in spacing-velocity plot. Extending the Newell's simplified car-following model, we finally discuss the distribution of points in the flow-density plot.

- **A traffic flow breakdown model for highway ramp bottlenecks: empirical observations, queueing theory and phase diagram analysis.**
  We will incorporate the queueing theory to describe traffic breakdown phenomena caused by ramping vehicle perturbations. The traffic breakdown probability directly corresponds with the probability that this jam queue dissipates in a given time period. The proposed queueing theory based model emphasizes the size evolution of a jam queue (local congested vehicle cluster) instead of its spatial evolutions. This model will capture the stochastic nature of traffic flow dynamics and therefore accounts for the probability of breakdown phenomena.
  We will propose a simple spatial-temporal queueing model based on Newell's simplified car-following model to quantitatively address some typical congestion patterns that were observed around on/off-ramps. Particularly, we examine three prime factors that play important roles in ramping traffic scenarios: the time for a vehicle to join a jam queue, the time for this vehicle to depart from this jam queue, and the time interval for the ramping vehicle to merge into the mainline. The analytically derived phase diagram will be compared with the simulation results. We will show that the new queueing model not only reserves the merits of Newell's model on the microscopic level but also helps quantify the contributions of these parameters in characterizing macroscopic congestion patterns.

In summary, from an academic point of view, our study advances the knowledge on stochastic traffic flow modeling in both microscopic car-following models and macroscopic fundamental diagram contexts; from a practice point of view, this study reveals the numerous underlying aspects of empirical traffic measurements, including the complex, dynamic and stochastic phenomena of highway traffic flow.

1.4 Organization

This book is organized as follows.

Chapter 2 reviews the development of traffic flow theory from the perspectives of macroscopic, mesoscopic, microscopic and stochastic approaches. Some representative models and methods in literature are summarized and compared. Particularly, various categories of headway/spacing distributions are investigated, such as univariate simple distribution, compositional distribution, mixed distribution and random matrix theory based approaches. The review summarizes four typical approaches in stochastic traffic flow modeling. The literature review forms a solid theoretical background for the book.

As shown in Fig. 1.1, we explore the characteristics of empirical traffic flow measurements to reveal the complex, dynamic and stochastic phenomena in Chap. 3. We apply the Eulerian and Lagrangian observations to qualitatively and quantitatively study some significant aspects of traffic flow features, including headway/spacing/velocity distributions, disturbances of congested platoons (jam queues) and time-frequency properties of oscillations. This work is the empirical foundation of the stochastic modeling.

In Chap. 4, from a microscopic perspective, we link two research directions of road traffic: mesoscopic headway distribution and microscopic vehicle interactions, by proposing a unified Markov state transition car-following model for different driving scenarios (highways and intersections), where random choice behaviors of headway/spacing (unconscious and inaccurate perceptions of time intervals and/or space) are the underlying stochastic sources. The parameters of this model can be directly estimated from NGSIM trajectory data. We also propose an asymmetric stochastic extension of the well-known Tau Theory by assuming that the observed headway distributions come from drivers consistent actions of headway adjusting, the intensity of headway change is proportional to the magnitude of headway on average.

The scattering feature of points on the density-flow plot remains as an attractive topic in the last few decades. In Chap. 5, from a macroscopic perspective, we discuss the implicit but tight connection between the microscopic spacing/headway distributions and the macroscopic scattering feature of flow-density plot. Although the scattering feature is influenced by various factors (e.g., driver heterogeneity and lane-changing/merging behaviors), we believe that may be mainly dominated by the microscopic spacing/headway distributions. Further extending the conventional deterministic reciprocal relations between flow rate and headway, we assume that the reciprocal of average headway of a homogeneous platoon and the corresponding flow rate should follow the same distribution. We examine microscopic driving behaviors that were retrieved from the NGSIM trajectory database. Results show that asymmetric driving behaviors result in a family of velocity-dependent lognormal type spacing/headway distributions. Then, we propose a stochastic platoon model to characterize the distribution of points in spacing- velocity plot. Tests on Performance Measurement System (PeMS) data reveal that the seemingly disorderly

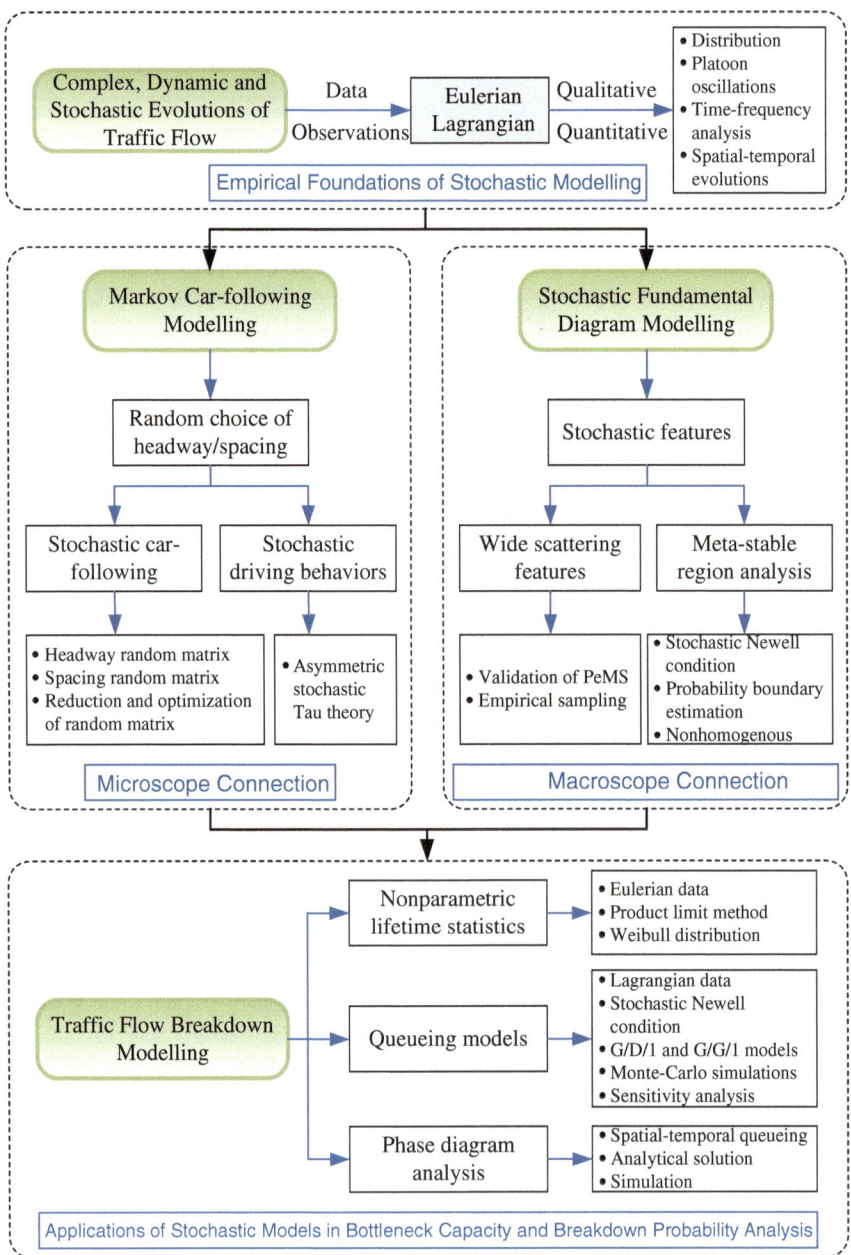

Fig. 1.1 Structure of the stochastic traffic flow modeling and applications

scattering points in the macroscopic flow-density plot follow the estimated flow rate distributions when the aggregation time interval is small enough (e.g., 30 s). If the aggregation time interval increases (e.g., to 5 min), the measured vehicles probably pass the loop detectors at different velocities and form heterogeneous platoons. It becomes difficult to find a definite distribution model that can fit the distribution of average headway/spacing for various heterogeneous platoons. However, most points in flow-velocity plot still locate within a certain 2D region, whose boundaries can be obtained from homogeneous platoon model. Finally, tests on PeMS verify this boundary property.

In Chap. 6, we study the commonly observed traffic flow breakdown phenomena at highway bottlenecks, emphasizing the transition process from perturbations to traffic jams in congested flow. We propose a queueing model to describe traffic breakdown phenomena caused by perturbations of on-ramp vehicles. If this jam queue cannot dissipate before the next vehicle merges into the main road, it often grows into a wide jam and results in traffic breakdown finally. In other words, the traffic breakdown probability directly corresponds with the probability that this jam queue dissipates in a given time period. But different from many previous models that focused on the propagation of jam waves, the proposed queueing theory based model emphasizes the size evolution of a jam queue (local congested vehicle cluster) instead of its spatial evolutions. We discuss the close relationship between TF-BP and the jam wave propagation phenomena with concentration on the evolution of jam queue size. Furthermore, a simplified queuing models are proposed based on Newells simplified car-following model, where the formation and dissipation of jam queues are treated as comprehensive effects of stochastic joining and leaving processes for upstream and downstream vehicles, respectively. Test results show that the simulated breakdown probability curve fits well with the empirically observed Weibull distribution type breakdown probabilities. This agreement indicates that this new model captures the stochastic nature of traffic flow dynamics and therefore accounts for the probability of breakdown phenomena. Finally, we analytically derive the phase diagram that is useful for presenting the evolution process for different traffic flow phases and quantitatively determining the phase transition conditions.

Chapter 7 summarizes this book and recommends a few future research directions.

Chapter 2
Literature Review

2.1 Introduction

This chapter briefly reviews the development of traffic flow theory from the perspectives of macroscopic, mesoscopic, microscopic, and stochastic approaches. Some representative models and methods are summarized. Particularly, characteristics of the stochastic traffic flow modeling will be emphasized and analyzed in detail. Since headway/spacing are the fundamental parameters highly correlated with traffic flow rate, density and speed, then probabilistic modeling of headway/spacing distributions will be categorized such as univariate distribution, compositional distribution, mixed distribution, and random matrix theory based approach. The joint distribution of headway/spacing/velocity can be regarded as a basis for stochastic modeling of traffic state evolutions. The historical development of traffic flow theory forms a solid theoretical background for the whole book.

2.2 Historical Development of Traffic Flow Theory

2.2.1 Macroscopic Modeling

The study of macroscopic continuum traffic flows began with the well-known Lighthill-Whitham-Richards (LWR) model or kinematic wave model which was proposed independently by Lighthill and Whitham (1955); Richards (1956). The model assumes that a discrete flow of vehicles can be approximated by a continuous flow. And then, vehicle dynamics can be described by the spatial vehicle density $\rho(x, t)$ as a function of location x and time t. As a result, many theoretical and numerical methods were developed to study this property based on the hyperbolic partial differential equation (PDE) type traffic flow model.

The conservation law of vehicles in an arbitrary stretch $[x_1, x_2]$ of the road, over an arbitrary time interval $[t_1, t_2]$, is written as:

© Tsinghua University Press, Beijing and Springer-Verlag Berlin Heidelberg 2015
X. (M.) Chen et al., *Stochastic Evolutions of Dynamic Traffic Flow*,
DOI 10.1007/978-3-662-44572-3_2

$$\int\limits_{x_1}^{x_2} \rho(x, t_2)dx - \int\limits_{x_1}^{x_2} \rho(x, t_1)dx = \int\limits_{t_1}^{t_2} \rho(x_1, t)v(x_1, t)dt - \int\limits_{t_1}^{t_2} \rho(x_2, t)v(x_2, t)dt \quad (2.1)$$

where $\rho(x, t)$ (veh/km) is the mean density at location x and time t, $v(x, t)$ (km/h) is the mean velocity at location x and time t.

However, weak solutions of Eq. (2.1) are not unique, and not all weak solutions capture the physics of traffic flow correctly (Jabari and Liu 2012). In order to derive the conservation law of traffic flow, Prigogine and Herman (1971) defined the phase space density as a production of $\rho(x, t)$ and the velocity distribution $\tilde{P}(v|x, t)$, i.e.,

$$\tilde{\rho}(x, t, v) \triangleq \rho(x, t)\tilde{P}(v|x, t) \quad (2.2)$$

where $\tilde{\rho}(x, t, v)$ is the phase space density function, that is, the joint PDF with respect to v at location x and time t, the velocity distribution satisfies $\int_0^\infty \tilde{P}(v|x, t)dv=1$, $\int_0^\infty v\tilde{P}(v|x, t)dv = v(x, t)$.

Let $\{(x, v) \mapsto (y, u)\}$ denotes the transition rate from state (x, v) to (y, u), where x and y represent locations, v and u represent velocities. Then the backward Kolmogorov type master equation that depicts the evolution of phase space density is

$$\frac{d\tilde{\rho}(x, t, v)}{dt} = \int\limits_0^\infty \int\limits_{-\infty}^\infty \{(y, u) \mapsto (x, v)\}\tilde{\rho}(y, t, u)dydu$$

$$- \int\limits_0^\infty \int\limits_{-\infty}^\infty \{(x, v) \mapsto (y, u)\}\tilde{\rho}(x, t, v)dydu \quad (2.3)$$

Suppose when $t \rightarrow \infty$, the phase space density tends to a steady state, i.e.,

$$\lim_{t \to \infty} \frac{d\tilde{\rho}(x, t, v)}{dt} = 0 \quad (2.4)$$

Expand Eq. (2.4) to the total derivative

$$\frac{\partial \tilde{\rho}(x, t, v)}{\partial t} + \frac{dx}{dt}\frac{\partial \tilde{\rho}(x, t, v)}{\partial x} + \frac{dv}{dt}\frac{\partial \tilde{\rho}(x, t, v)}{\partial v} = 0 \quad (2.5)$$

Integrate Eq. (2.5) with respect to v, we have

$$\int\limits_0^\infty \frac{\partial \tilde{\rho}(x, t, v)}{\partial t}dv + \frac{\partial \rho(x, t)}{\partial x}\int\limits_0^\infty v\tilde{P}(v|x, t)dv + \rho(x, t)\int\limits_0^\infty \frac{\partial \tilde{P}(v|x, t)}{\partial v}\frac{dv}{dt}dv = 0$$

$$(2.6)$$

Simplify Eq. (2.5), the reduced form is

$$\frac{\partial \rho(x,t)}{\partial t} + \frac{\partial \rho(x,t)}{\partial x} v(x,t) + \frac{\partial v(x,t)}{\partial x} \rho(x,t) = 0 \qquad (2.7)$$

or

$$\frac{\partial \rho(x,t)}{\partial t} + \frac{\partial \rho(x,t)v(x,t)}{\partial x} = 0 \qquad (2.8)$$

Incorporate the equilibrium flow function $q_e(\rho)$, we have the hyperbolic PDE of LWR model as

$$\frac{\partial \rho(x,t)}{\partial t} + \frac{\mathrm{d}q_e(\rho)}{\mathrm{d}\rho} \frac{\partial \rho(x,t)}{\partial x} = 0 \qquad (2.9)$$

The cell transmission model (CTM) was proposed by Daganzo (1994, 1995a) as a direct discretization of LWR model to simulate traffic flow evolutions using the Godunov Scheme (Lebacque 1996), in which the flow rate was modeled as a function of density with a triangular or trapezoidal form. Various modifications of the CTM model had been proposed in last two decades. For example, CTM was extended to model network traffic flow with general fundamental diagrams (Daganzo 1995b). Lags were introduced to formulate the lagged cell transmission model (LCTM) that adopted a nonconcave fundamental diagram, in the fact that, the forward wave speed was larger than the backward wave speed (Daganzo 1999). Recently, the original LCTM was modified by Szeto (2008) as the enhanced LCTM (ELCTM) to guarantee that the nonnegative densities would not be greater than the jam density. To validate the parameters online by loop detectors data, a switching mode model (SMM) was formulated, in which the evolution of traffic density switched among different sets of linear difference equations (Muñoz et al. 2003, 2006). The asymmetric cell transmission model (ACTM) was applied in optimal freeway ramp metering by Gomes and Horowitz (2006); Gomes et al. (2008). The cell-based dynamic traffic assignment formulation was further developed for networks. Lo et al. (2001); Lo and Szeto (2002); Boel and Mihaylova (2006) proposed the compositional CTM afterwards.

To model the evolutions of velocity more accurately, higher-order terms of density and velocity were incorporated. The higher-order density gradient type dynamic equation has the following expression

$$\frac{\partial v}{\partial t} + v\frac{\partial v}{\partial x} - \frac{v_e(\rho) - v}{\tau} - \frac{c_0^2}{\rho}\frac{\partial \rho}{\partial x} \qquad (2.10)$$

where $v = v(x,t)$, $\rho = \rho(x,t)$, $v_e(\rho)$ is the equilibrium velocity-density function, τ is the relaxation time, $c_0 \geqslant 0$ is the substitution variable with the same unit as velocity, $\{v + c_0, v - c_0\}$ are the characteristic velocities. $v + c_0$ is larger than the macroscopic velocity of traffic flow, so that this model was criticized by Daganzo (1995c).

In order to overcome the problem, higher-order velocity gradient type dynamic equation was derived as

$$\frac{\partial v}{\partial t} + v\frac{\partial v}{\partial x} = \frac{v_e(\rho) - v}{\tau} + c_0\frac{\partial v}{\partial x} \tag{2.11}$$

where characteristic velocities are not larger than the macroscopic velocity of traffic flow, i.e., $\{v, \ v - c_0\} \leqslant v$, so there are no reverse movements (Daganzo 1995c).

Helbing et al. (2009) proposed a general form for the higher-order model. They regarded that it was not contradictory when the characteristic velocity was larger than the macroscopic velocity. The general form is

$$\frac{\partial v}{\partial t} + v\frac{\partial v}{\partial x} = \frac{v_e(\rho) - v}{\tau} - \frac{1}{\rho}\left(\frac{\partial P_1}{\partial \rho}\frac{\partial \rho}{\partial x} + \frac{\partial P_2}{\partial v}\frac{\partial v}{\partial x}\right) \tag{2.12}$$

where $P_1 = P_1(\rho, v)$ and $P_2 = P_2(\rho, v)$ are pressure terms, satisfying $\partial_\rho P_1 \leqslant 0$, $\partial_v P_2 \leqslant 0$.

Analogously, we propose the following general macroscopic traffic flow model

$$\frac{\partial v}{\partial t} + v\frac{\partial v}{\partial x} = \frac{v_e(\rho) - v}{\tau} + k_1 v_e'(\rho)\frac{\xi}{\tau}\frac{\partial \rho}{\partial x} + k_2\frac{\eta}{\tau}\frac{\partial v}{\partial x} \tag{2.13}$$

where k_1, k_2 are the weighted coefficients for the density gradient term and the velocity gradient term, $\xi = \xi(\rho, v)$ and $\eta = \eta(\rho, v)$ reflect the anticipative and adaptive driving behaviors, respectively.

We obtain the general simultaneous PDEs

$$\frac{\partial \rho}{\partial t} + v_e(\rho)\frac{\partial \rho}{\partial x} + \rho\frac{\partial v}{\partial x} = 0 \tag{2.14}$$

$$\frac{\partial v}{\partial x} + v\frac{\partial v}{\partial t} = \frac{v_e(\rho) - v}{\tau} + k_1 v'_e(\rho)\frac{\xi}{\tau}\frac{\partial \rho}{\partial x} + k_2\frac{\eta}{\tau}\frac{\partial v}{\partial x} \tag{2.15}$$

The analytical linear stability condition for this general model is as following (refer to the detailed derivation in Appendix A)

$$\rho_0 v_e'(\rho_0) + k_1\frac{\xi}{\tau} + k_2\frac{\eta}{\tau} \geqslant 0 \tag{2.16}$$

Characteristic velocities are $v + c_\pm$, where

$$c_\pm = -\frac{|\eta|}{2\tau} \mp \sqrt{\rho v_e'(\rho)\frac{\xi}{\tau} + \left(\frac{\eta}{2\tau}\right)^2} \tag{2.17}$$

2.2.2 Mesoscopic Modeling

Common mesoscopic models belong to three categories (Hoogendoorn and Bovy 2001): time-headway distribution models, cluster models, and gas-kinetic models.

Table 2.1 Typical macroscopic traffic flow models

Models	k_1	k_2	ξ	η	Characteristic velocities $v + c_{\pm}$
Payne (1971)	1	0	$\frac{1}{2\rho}$	–	$c_{\pm} = \mp\sqrt{\frac{\|v'_e(\rho)\|}{2\tau}}$
Whitham (1974)	1	0	$-\frac{\mu}{\rho v'_e(\rho)}$	–	$c_{\pm} = \mp\sqrt{\mu/\tau}$
Phillips (1979)	1	0	$\frac{\tau\Theta_0(1-\rho/\rho_j)}{\rho v'_e(\rho)}$	–	$c_{\pm} = \mp\sqrt{\Theta_0\left(1-\rho/\rho_j\right)}$
Zhang (1998)	1	0	$-\rho v'_e(\rho)\tau$	–	$c_{\pm} = \pm\rho v'_e(\rho)$
Aw and Rascle (2000) Greenberg (2001)	0	1	–	$\tau\gamma\rho^{\gamma}$	$c_{+} = -\gamma\rho^{\gamma} \leqslant 0, \ c_{-} = 0$
Zhang (2002)	0	1	–	$-c(\rho)\tau$	$c_{+} = c(\rho) \leqslant 0, \ c_{-} = 0$
Jiang et al. (2001, 2002)	0	1	–	$c_0\tau$	$c_{+} = c_0 \leqslant 0, \ c_{-} = 0$
Xue and Dai (2003)	0	1	–	$-t_r\rho v'_e(\rho)$	$c_{+} = \frac{t_r}{\tau}\rho v'_e(\rho) \leqslant 0, \ c_{-} = 0$
Helbing and Johansson (2009)	1	1	$\frac{\tau\partial_\rho P_1}{v'_e(\rho)\rho}$	$-\frac{\tau\partial_v P_2}{\rho}$	$c_{\pm} = \frac{\partial_v P_2}{2\rho} \mp \sqrt{\partial_\rho P_1 + (\frac{\partial_v P_2}{2\rho})^2}$

The book will discuss headway distribution models in Sect. 2.3. Due to the space limitation, cluster models will be omitted. We will briefly show the idea of gas-kinetic model (Table 2.1). Prigogine and Herman (1971) proposed the following Boltzmann equation

$$\frac{d_v\tilde{\rho}}{dt} = \frac{d\tilde{\rho}}{dt} + v\frac{d\tilde{\rho}}{dx} = \left(\frac{d\tilde{\rho}}{dt}\right)_{acc} + \left(\frac{d\tilde{\rho}}{dt}\right)_{int} \tag{2.18}$$

Acceleration behaviors can be modeled by the relaxation process that transforms from velocity distribution $\tilde{P}(v; x, t)$ to the expected velocity distribution $\tilde{P}_0(v)$

$$\left(\frac{d\tilde{\rho}}{dt}\right)_{acc} = \frac{\rho(x,t)}{\tau(\rho(x,t))}\left[\tilde{P}_0(v) - \tilde{P}(v|x,t)\right] \tag{2.19}$$

Interactions among vehicles are

$$\left(\frac{d\tilde{\rho}}{d}t\right)_{int} = (1 - p(\rho))\rho(x,t)\left[\bar{v}(x,t) - v\right]\tilde{\rho}(x,t,v) \tag{2.20}$$

where

$$\bar{v}(x,t) = \int_0^\infty v\tilde{P}(v|x,t)dv = \int_0^\infty v\frac{\tilde{\rho}(x,t,v)}{\rho(x,t)}dv \tag{2.21}$$

Recently, mesoscopic traffic simulation has been attracted more efforts on the operations of dynamic traffic systems, e.g., CONTRAM (Leonard et al. 1989), DYNASMART (Jayakrishnan et al. 1994), FASTLANE (Gawron 1998), DYNAMIT (Ben-Akiva et al. 2010), INTEGRATION (van Aerde and Rakha 2002), MEZZO

(Burghout et al. 2006), DYNAMEQ (Snelder 2009), dynaCHINA (Lin and Song 2006), etc.

2.2.3 Microscopic Modeling

Microscopic traffic flow model utilizes the *Lagrangian* method to study traffic flow dynamics by describing one vehicular trajectory or interactions among multiple vehicles. Microscopic modeling can be divided into car-following model and lane changing model. This section only reviews the theoretical development of car-following models. Common car-following models include: stimulus response model, safe distance or behavioral model, psychological-physical/action point model, artificial intelligence-based model, cellular automaton (CA), etc.

The updating equations of velocity and location are

$$\begin{cases} v_n(t + \Delta t) = v_n(t) + \dot{v}_n(t)\Delta t \\ x_n(t + \Delta t) = x_n(t) + v_n(t)\Delta t + \frac{1}{2}\dot{v}_n(t)(\Delta t)^2 \end{cases} \tag{2.22}$$

where $x_n(t)$ is the location of the nth vehicle at time t, $v_n(t) = \dot{x}_n(t)$ is the velocity of the nth vehicle at time t, Δt is the updating time step.

The general form of acceleration equation is

$$a_n = f(x_{n-1}, x_n, v_{n-1}, v_n), \ n \in \mathbb{N}^+ \tag{2.23}$$

Table 2.2 shows the long-term evolution of typical microscopic car-following models. We can expand Eq. (2.23) to the scenario of a multiple-car-following model as

$$a_n = f(x_n, \ldots, x_{n-m+1}; v_n, \ldots, v_{n-m+1}), \ m, n \in \mathbb{N}^+ \tag{2.24}$$

where m is the number of vehicles that influence the nth vehicle.

In the past decade, some multi-anticipative car-following models were proposed to enhance the stability of traffic flow. One approach assumed that the individual location and velocity information could be shared among different vehicles to simulate multi-anticipative behaviors via inter-vehicle communications (Li and Wang 2007). Other approaches emphasized the multi-anticipative behaviors of human drivers and tried to model the actions of drivers via simulation models (Treiber et al. 2006a). Numerical experiments showed that multi-vehicle interactions generally enlarged the stable region of traffic flow. For example, based on the extended optimal velocity model (OVM) and full velocity difference model (FVDM), Lenz et al. (1999) showed that the stability of traffic flow was improved by taking into account relative velocities of vehicles. Results indicated that the multi-anticipative behavior enlarged the linear stability region. On the contrary, the human reaction or manipulation delays might lead to the instability of traffic flow. Appendix B applies perturbation

Table 2.2 Typical microscopic traffic flow models

Models	Acceleration equations [a]	Parameters
Pipes (1953)	$a_n(t + \tau) = c(v_n(t) - v_{n-1}(t))$	c
Gazis et al. (1961)	$a_n(t) = cv_n^m(t)\dfrac{v_n(t) - v_{n-1}(t)}{(x_n(t) - x_{n-1}(t))^l}$	c, m, l
Newell (1961)	$a_n(t) = c(x_n(t) - x_{n-1}(t))^l$	c
	$a(t) = \dfrac{1}{\tau}[v_{opt}(x_n(t) - x_{n-1}(t)) - v_{n-1}(t)]^b$	c, d
Bierley (1963)	$a_n(t) = \alpha(v_n(t) - v_{n-1}(t)) + \beta(x_n(t) - x_{n-1}(t))^l$	α, β
Sultan et al. (2004)	$a_n(t + \tau) = cv_n^m(t)\dfrac{\Delta v_n(t)}{\Delta x_n^l(t)} + k_1 a_{n-1}(t) + k_2 a_n(t)$	c, k_1, k_2
Bando et al. (1995)	$a_n(t) = c[v_{opt}(x_n(t) - x_{n-1}(t)) - v_n(t)]^c$	c, s_{safe}
Helbing and Tilch (1998)	$a_n(t) = c[v_{opt}(\Delta x_n(t)) - v_n(t)] + \lambda H \Delta v_n(t)^d$	c, λ
Jiang et al. (2001)	$a_n(t) = c[v_{opt}(\Delta x_n(t)) - v_n(t)] + \lambda \Delta v_n(t)$	c, λ
Treiber et al. (2000)	$a_n(t) = a_{max}\left[1 - \left(\dfrac{v_n}{v_{max}}\right)^\delta - \left(\dfrac{s^*(v_n, \Delta v_n)}{s_n}\right)^2\right]^e$	δ, T_n, s_n

[a] $\Delta x_n(t) - \Delta x_{n,n-1}(t) - x_n(t) - x_{n-1}(t)$, $\Delta v_n(t) = \Delta v_{n,n-1}(t) = v_n(t) - v_{n-1}(t)$
[b] $v_{opt}(x_n(t) - x_{n-1}(t)) = v_{free}\left[1 - \exp\left(-\dfrac{c}{v_{free}}(x_n(t) - x_{n-1}(t)) - d\right)\right]$
[c] $v_{opt}(x_n(t) - x_{n-1}(t)) = \dfrac{v_{free}}{2}\left[\tanh(X_n(t) - x_{n-1}(t) - S_{safe}) + \tanh(S_{safe})\right]$
[d] $H = H(-\Delta v_n(t))$ is Heaviside function
[e] $s^*(v_n, \Delta v_n) = s_{min} + \max\{T_n v_n + v_n \Delta v_n/(2\sqrt{a_{max} b_n}), 0\}$

analysis (PA) to derive the critical linear stability condition for multi-car-following models.

However, the common problem of typical models listed in Table 2.2 is to define a deterministic acceleration equation. In field applications, vehicles are influenced by many stochastic internal and external factors that are not taken into consideration in the classic deterministic acceleration equation based models. How to depict the stochastic characteristics of driving behaviors and time-varying traffic flow states more accurately will be discussed in Chap. 4 by using the Markov model based on headway/spacing distributions.

2.2.4 Stochastic Modeling

Road traffic flow is influenced by various random factors, including both external factors such as the weather, and internal factors such as transportation facilities, vehicle characteristics, driver behaviors, etc. These stochastic factors make the deterministic approaches difficult to accurately estimate or predict dynamic traffic evolutions. To overcome this problem, numerous stochastic approaches were developed for continuous traffic flow modeling. In this study, they are divided into the following four categories:

- Macroscopic traffic flow modeling by the randomization of the first-order conservation law and/or higher-order momentum equations;

- Microscopic traffic flow modeling by the randomization of driving behaviors and/or mixed traffic flows;
- Fundamental diagram and the corresponding phase transitions by the randomization of relationships among flow, density, and velocity;
- Transportation reliability studies on the randomization of road capacity and/or travel time distribution.

In summary, the four categories of stochastic approaches can be generally written as

$$\Theta(x + \Delta x, t + \Delta t) = f(\Theta(x, t), \Delta x, \Delta t) + \epsilon(x, t) \qquad (2.25)$$

where $\Theta(x, t)$ is the vector of traffic states at location x and time t, $f(\cdot)$ is the traffic state evolution function, $\epsilon(x, t)$ is noise function.

First, in macroscopic modeling, Boel and Mihaylova (2006) proposed a stochastic compositional model for freeway traffic flows, where the randomness was reflected in the probability distributions of sending and receiving functions, also in the well-defined noise term of speed adaptation rules. Sumalee et al. (2011) proposed a first-order macroscopic stochastic cell transmission model (SCTM), each operational mode of which was formulated as a discrete time bilinear stochastic system to model traffic density of freeway segments in stochastic demand and supply. However, this approach still assumed a deterministic FD with a second-order wide-sense stationary (WSS) noisy disturbance. Wang et al. (2005, 2006, 2009b) presented a general stochastic macroscopic traffic flow model of freeway stretches based on a traffic state estimator using extended Kalman filtering and developed the freeway network state monitoring software (i.e., REal-time motorway Network trAffIc State Surveil-lANCE, RENAISSANCE).

Secondly, in microscopic modeling, Wagner (2011) proposed a time-discrete stochastic harmonic oscillator for car-following based on the deterministic acceleration Eq. (2.23), i.e., $a_n = f(x_{n-1}, x_n, v_{n-1}, v_n) + \epsilon$, $n \in \mathbb{N}^+$ where ϵ is the noise term. Huang et al. (2001) proposed a stochastic CA model by incorporating braking probability, occurrence, and dissipation probability. Some other approaches include Nagel and Schreckenberg (1992); Zhu et al. (2007). Since microscopic car-following model is highly correlated with headway/spacing distributions, we will further discuss a Markov model to depict headway/spacing evolutions in Chap. 4.

Thirdly, in stochastic FD and its phase transition analysis, the classic assumption is the existence of deterministic functions of flow-density and speed-density. FD has been the foundation of traffic flow theory and transportation engineering. According to the definition by Edie (1961), in the $t \sim x$ vehicular trajectory diagram, we have

$$\rho = \sum_{i=1}^{n} \frac{T_i}{|A|}, \quad q = \sum_{i=1}^{n} \frac{D_i}{|A|}, \quad v = \frac{q}{\rho} = \frac{\sum_{i=1}^{n} D_i}{\sum_{i=1}^{n} T_i} \qquad (2.26)$$

where $|A|$ is the area of an arbitrary region A, T_i, and D_i are the travel time and distance for the ith vehicle in A.

Treiber et al. (2006b) investigated the adaptation of headways in car-following models as a function of the local velocity variance to study the scattering features in flow-density plot. Ngoduy (2011) argued that the widely scattering flow-density relationship might be caused by the random variations in driving behavior. The distribution features and probabilistic boundaries estimation method will be further discussed in Chap. 5.

At last, in transportation reliability studies, Brilon et al. (2005) pointed out that the concept of stochastic capacities was more realistic and more useful than the traditional concept of deterministic capacity. In the last decade, many efforts were made to identify the characteristics of traffic flow breakdown and its occurrence condition (Evans et al. 2001; Kerner and Klenov 2006; Kesting et al. 2010; Smilowitz and Daganzo 2002). Usually, traffic breakdown phenomena can be triggered by external disturbances or internal perturbations. This book only considers the latter one that has been widely observed and validated when studying the features of oscillations (Banks 2006; Del Castillo 2001; Jost and Nagel 2003; Kerner and Klenov 2006; Kim and Zhang 2008; Lu and Skabardonis 2007; Son et al. 2004; Wang et al. 2007). Many approaches can be used in traffic flow breakdown phenomena. For example, Bassan et al. (2006) used the mathematical property of log periodic oscillations (LPO) to model traffic density over time. Habib-Mattar et al. (2009) developed a density-versus-time model to describe traffic breakdown, it was found that density increased sharply toward the peak period and then decreased and increased again toward the breakdown. Since density cannot be directly measured in field, many researchers tend to study the relationship between traffic flow breakdown probability with the upstream flow. In general, the probability shows an increasing sigmoid curve in terms of the upstream flow, where Weibull distribution is commonly incorporated to formulate the curve (Banks 2006; Brilon et al. 2005; Chow et al. 2009; Lorenz and Elefteriadou 2001; Mahnke and Kühne 2007). Chen and Zhou (2010) proposed the α-reliable mean-excess traffic equilibrium (METE) model that explicitly considered both reliability and unreliability aspects of travel time variability in the route choice decision process. Stochastic capacity is highly correlated with traffic flow breakdown probability, and the analytical derivation of phase transition will be present in Chap. 6.

Differing from the above four kinds of stochastic approaches, Mahnke et al. (2001, 2005); Mahnke and Kühne (2007); Mahnke and Pieret (1997) applied stochastic process to the dynamic mechanism of traffic congestion occurrence and dissipation based on time-varying probability distributions, by using master equation of statistical physics to analyze the phase transitions and nucleation phenomena in jam queues. The general form of master equation is (Chowdhury et al. 2010; van Kampen 2007)

$$\frac{\partial \mathcal{P}(z, t)}{\partial t} = \int [\omega_{z'z}\mathcal{P}(z', t) - \omega_{zz'}\mathcal{P}(z, t)]dz' \qquad (2.27)$$

where z and z' are continuous state variables, t is continuous time, $\mathcal{P}(z, t)$ is the probability of being in state z at time t, $\omega_{zz'}$ is the transition rate from state z to z', satisfying $\int \omega_{zz'}dz' = \int \omega_{z'z}dz' = 1$.

In stochastic traffic flow modeling, traffic states are usually represented by discrete variables. Let S be an arbitrary discrete state, the probability that traffic belongs to state S at time t is $\mathcal{P}_S(t)$. The evolution of traffic states can be described by the following discrete master equation

$$\frac{d\mathcal{P}_S(t)}{dt} = \sum_{S'} \omega_{S'S} \mathcal{P}_{S'}(t) - \sum_{S'} \omega_{SS'} \mathcal{P}_S(t) \qquad (2.28)$$

where $\omega_{SS'}$ is the transition rate from state S to S', satisfying $\sum_{S'} \omega_{SS'} = \sum_{S'} \omega_{S'S} = 1$, the first term on the right side is the transition rate from one state S' to the current state S, the second term is the transition rate from the current state S to another state S'.

Furthermore, when traffic state and time are both discrete, suppose traffic state is S at time t, according to the master equation in Eq. (2.28), at time $t + \Delta t$, the probability that traffic still belongs to state S is

$$\mathcal{P}_S(t + \Delta t) = \left(1 - \sum_{S' \neq S} \omega_{SS'} \Delta t\right) \mathcal{P}_S(t) + \sum_{S' \neq S} \omega_{S'S} \Delta t \mathcal{P}_{S'}(t) \qquad (2.29)$$

Regard the formation and dissipation of a jam queue as a Markov process, and assume at most one vehicle can join the jam queue in a differentiation δt, i.e., the probability that two or more vehicles join the jam queue in δt is $o(\delta t)$, then the master equation of the jam queue length distribution is

$$\frac{\partial \mathcal{P}(n, t)}{\partial t} = \omega_+(n - 1)\mathcal{P}(n - 1, t) + \omega_-(n + 1)\mathcal{P}(n + 1, t)$$
$$- [\omega_+(n)\mathcal{P}(n, t) + \omega_-(n)\mathcal{P}(n, t)], \ n \in \mathbb{N}^+ \qquad (2.30)$$

where $\omega_+(n)$ and $\omega_-(n)$ are the transition rates for a jam queue length changes of $\{n \mapsto n + 1\}$ and $\{n \mapsto n - 1\}$ vehicles, respectively.

Mahnke et al. (2001, 2005); Mahnke and Kühne (2007); Mahnke and Pieret (1997) studied the occurrence and evolution of jam queues in a homogeneous circle road with periodic boundary conditions, and defined the joining and leaving rates as

$$\omega_+(n) = \frac{v_{opt}(\Delta x_{free}(n)) - v_{opt}(\Delta x_{cong}(n))}{\Delta x_{free}(n)) - \Delta x_{cong}(n)}, \quad \omega_-(n) = \frac{1}{\tau_{out}} \qquad (2.31)$$

where the joining rate is calculated by the OVM, the optimal free-flow velocity is $v_{opt}(\Delta x_{free}(n))$, the congested optimal velocity is $v_{opt}(\Delta x_{cong}(n))$, and $\Delta x_{cong}(n) = \Delta x_{cong}$. Then we get the jam queue length $L_{cong} = n \Delta x_{cong}$ and the free-flow length $L_{free} = L - L_{cong} = n \Delta x_{cong}$, L is the length of the circle road. τ_{out} is the mean waiting time for a vehicle to leaving a jam queue.

Furthermore, Kühne et al. (2002) applied the master equation-based nucleation model to traffic flow breakdown phenomenon in ramping bottlenecks, and formulated the Fokker-Planck equation that described the jam queue evolutions.

2.3 Probabilistic Headway/Spacing Distributions

2.3.1 Simple Univariable Distributions

• **Negative exponential distribution**

Let h denote the random variable of headway. Negative exponential distribution describes the interarrival time as Poisson process. Events occur continuously and independently at a constant average rate. It is appropriate when traffic flow rate and density are small.

The PDF is

$$f(h|\lambda) = \lambda e^{-\lambda h}, \ h \geqslant 0 \tag{2.32}$$

where λ is the rate parameter. It indicates aggressive driving behaviors when λ is relatively large, while it indicates timid driving behaviors when λ is relatively small.

The CDF is

$$F(h|\lambda) = 1 - e^{-\lambda h}, \ h \geqslant 0 \tag{2.33}$$

The expectation and variance of headway are

$$E[h] = \frac{1}{\lambda}, \ \text{Var}[h] = \frac{1}{\lambda^2} \tag{2.34}$$

Define the maximum likelihood function as

$$\ell(\lambda) \triangleq \lambda^n \exp\left(-\lambda \sum_{i=1}^n h_i\right) \tag{2.35}$$

Let $\partial \ell(\lambda)/\partial \lambda = 0$, we have the following maximum likelihood estimator (MLE)

$$\hat{\lambda} = \frac{1}{\bar{h}} \tag{2.36}$$

where $\bar{h} = \frac{1}{n}\sum_{i=1}^n h_i$, $\{h_1, \ldots, h_n\}$ are samples.

- **Shifted exponential distribution**

Since the probability of headway near zero is large in negative exponential distribution, to avoid the extremely short headways, the PDF is shifted rightwards with a deterministic positive value.

The PDF is

$$f(h|\lambda, h_0) = \lambda e^{-\lambda(h-h_0)}, \ h \geqslant h_0 \tag{2.37}$$

where h_0 is the translation parameter.

The CDF is

$$F(h|\lambda, h_0) = 1 - e^{-\lambda(h-h_0)}, \ h \geqslant h_0 \tag{2.38}$$

The expectation and variance of headway are

$$E[h] = \frac{1}{\lambda} + h_0, \ \text{Var}[h] = \frac{1}{\lambda^2} \tag{2.39}$$

Define the maximum likelihood function as

$$\ell(\lambda, h_0) \triangleq \lambda^n \exp\left(-\lambda \sum_{i=1}^{n} (h_i - h_0)\right) \tag{2.40}$$

Let $\partial \ell(\lambda, h_0)/\partial \lambda = 0$, we have the MLEs

$$\hat{h}_0 = \min\{h_1, \ldots, h_n\}, \ \hat{\lambda} = \frac{1}{\bar{h} - \hat{h}_0} \tag{2.41}$$

- **Gamma distribution (Pearson type III distribution)**

Gamma distribution is a two-parameter family of continuous probability distributions used to model headway commonly.

The PDF is

$$f(h|\alpha, \beta, h_0) = \frac{(h - h_0)^{\alpha-1} e^{-(h-h_0)/\beta}}{\beta^\alpha \Gamma(\alpha)}, \ h \geqslant h_0 \tag{2.42}$$

where $\Gamma(\alpha) = \int_0^\infty t^{\alpha-1} e^t \, dt$ is the gamma function with a shape parameter α and a scale parameter β.

The CDF is

$$F(h|\lambda, h_0) = \frac{\gamma(\alpha, (h - h_0)/\beta)}{\Gamma(\alpha)}, \ h \geqslant h_0 \tag{2.43}$$

where $\gamma(\alpha, (h - h_0)/\beta) = \int_0^{(h-h_0)/\beta} t^{\alpha-1} e^{-t} \, dt$.

The expectation and variance are

$$E[h] = \alpha\beta + h_0, \ \ \text{Var}[h] = \alpha\beta^2 \tag{2.44}$$

There is no closed-form solution for α and β. They can be numerically approximated by using Newton's method, method of moments, etc. However, Gamma distribution cannot be suitable to depict headway distribution when the shape parameter is larger than 1 because the bell-like shape gives low probability to short headways.

- **Shifted lognormal distribution**

If $\log(h - h_0)$ follows the normal distribution $\mathcal{N}(\mu_h, \sigma_h^2)$, then h belongs to shifted lognormal distribution. The lognormal relation holds if the change in a headway during a small time interval is a random proportion of the headway at the start of the interval, and the mean and the variance of the headway remain constant over time (Luttinen 1996). Shifted lognormal distribution is widely used in headway/spacing modeling in both scenarios of continuous and interrupted transportation facilities. It is also closely related to car-following models as well.

The PDF is

$$f(h|\mu_h, \sigma_h, h_0) = \frac{1}{\sqrt{2\pi}\sigma_h(h - h_0)} \exp\left(-\frac{(\log(h - h_0) - \mu_h)^2}{2\sigma_h^2}\right), \ \ h \geqslant h_0 \tag{2.45}$$

where μ_h, σ_h are the mean and standard deviation, respectively, of the headway's natural logarithm

The CDF is

$$F(h|\mu_h, \sigma_h, h_0) = \Phi\left(\frac{\log(h - h_0) - \mu_h}{\sigma_h}\right), \ \ h \geqslant h_0 \tag{2.46}$$

where $\Phi(\cdot)$ is the standard normal distribution function.

The expectation and variance of headway are

$$E[h] = \exp\left(\mu_h + \sigma_h^2/2\right) + h_0, \ \ \text{Var}[h] = \exp\left(2\mu_h + \sigma_h^2\right)\left(\exp(\sigma_h^2) - 1\right) \tag{2.47}$$

Define the maximum likelihood function as

$$\ell(\mu_h, \sigma_h, h_0) \triangleq \prod_{i=1}^{n} f(h_i|\mu_h, \sigma_h, h_0) \tag{2.48}$$

Let

$$\frac{\partial \ell(\hat{\mu}_h, \hat{\sigma}_h, \hat{h}_0)}{\partial \hat{\mu}_h} = \frac{\partial \ell(\hat{\mu}_h, \hat{\sigma}_h, \hat{h}_0)}{\partial \hat{\sigma}_h} = \frac{\partial \ell(\hat{\mu}_h, \hat{\sigma}_h, \hat{h}_0)}{\partial \hat{h}_0} = 0$$

we have the following MLEs

$$\hat{h}_0 = \min\{h_1, \ldots, h_n\} \tag{2.49a}$$

$$\hat{\mu}_h = \frac{1}{n} \sum_{i=1}^{n} \log(h_i - \hat{h}_0) \tag{2.49b}$$

$$\hat{\sigma}_h = \frac{1}{n} \sum_{i=1}^{n} \left(\log(h_i - \hat{h}_0)\right)^2 - \left(\frac{1}{n} \sum_{i=1}^{n} \log(h_i - \hat{h}_0)\right)^2 \tag{2.49c}$$

2.3.2 Compositional Distributions

The simple univariable distributions are incapable of describing both sharp peak and long tail properties of headway/spacing. Because of the coexistence of two main traffic flow states, i.e., free-flow headways and car-following headways, whose distributions are significantly different. So it's better to incorporate the compositional distribution functions.

The compositional PDF of headway is defined as $f(h)$, i.e.,

$$f(h) = \varphi p(h) + (1 - \varphi)q(h) \tag{2.50}$$

where $0 \leqslant \varphi < 1$ is the proportion of constrained headways in car-following states, $p(h)$ is the PDF of constrained headways, $q(h)$ is the PDF of free-flow headways.

According to the convolution formula, we have

$$q(h) = p^*(\lambda)^{-1} \lambda e^{-\lambda h} \int_0^h p(z)dz \tag{2.51}$$

where $p^*(\lambda)$ is the Laplace transform of $p(h)$, i.e.

$$p^*(\lambda) = \int_0^\infty e^{-\lambda h} p(h)dh \tag{2.52}$$

• **Hyperexponential distribution**

Schuhl (1955) first applied Hyperexponential distribution to model headway, known as Schuhl's (composite exponential) distribution.

The PDFs are

$$p(h) = \lambda_p e^{-\lambda_p(h - h_0)}, \quad q(h) = \lambda_q e^{-\lambda_q(h - h_0)} \tag{2.53}$$

where λ_p, λ_q are parameters of exponential distributions for car-following and free-flow states, respectively.

- **Hyperlang distribution**

Dawson and Chimini (1968) suggested the Erlang-distribution as a model for car-following headways, i.e.,

$$p(h) = \frac{\lambda_p^k (h - h_0)^{k-1} e^{-\lambda_p(h - h_{0,p})}}{(k-1)!}, \quad q(h) = \lambda_q e^{-\lambda_q(h - h_{0,q})} \tag{2.54}$$

where $k \in \mathbb{N}^+$, λ_p, λ_q are parameters of car-following and free-flow distributions, respectively. $h_{0,p}$, $h_{0,q}$ are the translation parameters. Luttinen (1996) pointed out that the hyperlang distribution has an exponential tail, and the shape of the PDF is similar to empirical headway distributions.

2.3.3 Mixed Distributions

It is found that many stationary distribution models could fit the empirical data of free flow but not congested flow. One way to solve this problem is to use the mixed headway distribution models. For example, in the M3 model (Cowan 1975), the headways of free-driving vehicles and those leader-following vehicles were assumed to follow different PDFs. But the calibration of mixed models is usually tedious, if we want to fit the empirical distributions with a high accuracy.

- **Semi-Poisson distributions**

Buckley (1968) proposed the semi-Poisson model that described the fluctuations in car-following states, conjectured there was a zone of emptiness in front of each vehicle, and compared it with Gamma distribution, shifted Gamma distribution, exponential distribution by using field measurements. Wasielewski (1974) applied non-parametric method to calculate constrained headway in semi-Poisson distribution.

The PDF of semi-Poisson distribution is

$$f(h) = \varphi p(h) + (1 - \varphi) \frac{\int_0^h p(h)\mathrm{d}t}{p^*(\theta)} \theta e^{-\theta h}, \quad h \geqslant 0, \ \theta > 0 \tag{2.55}$$

where $p^*(\theta) = \int_0^\infty e^{-\theta h} p(h)\mathrm{d}h$ is the Laplace transform of $p(h)$.

Semi-Poisson distribution defines two common headway distributions for car-following scenarios, i.e.,

Gamma distribution and its Laplace transform

$$p(h) = \frac{h^{\alpha-1} e^{-h/\beta}}{\beta^\alpha \Gamma(\alpha)}, \quad q^*(\lambda) = (1 + \lambda\beta)^{-\alpha} \tag{2.56}$$

Gaussian distribution and its Laplace transform

$$p(h) = \frac{1}{\sqrt{2\pi}\,\sigma} e^{-\frac{(h-\mu)^2}{2\sigma^2}}, \quad p^*(\lambda) = e^{(\sigma^2/2-\mu)\lambda} \tag{2.57}$$

where μ and σ are the expectation and standard deviation of Gaussian distribution.

Branston (1976) proposed a mixed model of a generalized queuing model and semi-Poisson model. The PDF for free-flow headway is

$$q(h) = \lambda e^{-\lambda h} \int_0^h p(z) e^{\lambda z} \mathrm{d}z \tag{2.58}$$

where $1/\lambda$ is the average headway. Assume headway follows lognormal distribution in car-following mode as

$$p(h) = \frac{1}{\sqrt{2\pi}\,\sigma h} e^{-\frac{(\log h - \mu)^2}{2\sigma^2}} \tag{2.59}$$

then the PDF for mixed headway is

$$f(h) = \varphi p(h) + (1-\varphi)\lambda e^{-\lambda h} \int_0^h p(z) e^{\lambda z} \mathrm{d}z \tag{2.60}$$

2.3.4 Random Matrix Model

Krbálex and Šěba (2001) showed that traffic data from different sources belonged to a class of random matrix distributions. Abul-Magd (2006) applied the random matrix theory (RMT) to the car-parking problem and adopted a Coulomb gas model that associated coordinates of gas particles with the eigenvalues of a random matrix, in which the Wigner surmise for Gaussian unitary ensemble (GUE) was given by

$$P(\varsigma) = \frac{32}{\pi^2} \varsigma^2 e^{-4\varsigma^2/\pi}, \quad \varsigma \geqslant 0 \tag{2.61}$$

where ς is the space gap, i.e. spacing minus vehicle length.

Abul-Magd (2007) pointed out that RMT modeled the Hamiltonians of chaotic systems as members of an ensemble of random matrices that depended only on the symmetry properties of the system and GUE modeled systems violating time reversal symmetry. In the phase transition from the free-flow state to the congested state, the GUE of RMT is

$$P_{\mathrm{GUE}}(\varsigma) = \frac{32\varsigma^2}{\pi^2\bar{\varsigma}^3}e^{-4\varsigma^2/\pi\bar{\varsigma}^2} \tag{2.62}$$

where $\bar{\varsigma}$ is the mean space gap.

Space gaps in free-flow traffic are uncorrelated and follow the Poisson distribution

$$P_{\mathrm{Poisson}}(\varsigma) = \frac{1}{\bar{\varsigma}}e^{-\varsigma/\bar{\varsigma}} \tag{2.63}$$

More recently, models based on headway/spacing distributions and state transitions received more interests. For example, Jin et al. (2009) revealed that the distributions of departure headways at each position in a queue approximately followed a certain lognormal distribution except the first one by using video data collected from four intersections in Beijing. Wang et al. (2009a) estimated a cellular automation model for spacing distribution.

2.4 Summary

This chapter reviews the history of traffic flow theory from the perspectives of macroscopic, mesoscopic, microscopic modeling approaches, and summaries their scopes of applications. This chapter focuses on the stochastic traffic flow modeling method and the headway/spacing probabilistic distributions. Based on the overview of the state-of-the-art traffic flow model, the historical development of traffic flow modeling approaches are summarized as a literature foundation for the other chapters.

Chapter 3
Empirical Observations of Stochastic and Dynamic Evolutions of Traffic Flow

3.1 Introduction

Road traffic flow is characterized by the complex, dynamic, and stochastic phenomena. In history, traffic theorists and transportation engineers applied numerous approaches to the qualitative or quantitative analyses of traffic flow. Figure 3.1 shows the empirical investigations based on *Eulerian* and *Lagrangian* field measurements in this chapter. The studies of stochastic phenomena include headway/spacing/velocity distributions, disturbances of congested platoons (jam queues), and time-frequency properties of oscillations. These observations will reveal some important aspects of complex, dynamic, and stochastic evolutions of traffic flow.

The observations in this chapter are the empirical basis for the book. The relations with other parts in the book are: (1) the studies of empirical joint distributions of headway/spacing/velocity based on typical *Lagrangian* measurements, i.e., trajectory data, promise traffic state transition matrices for a Markov car-following model that will be further discussed in Chap. 4; (2) based on the *Lagrangian* measurements, the spatial-temporal evolutions of mean velocity, and disturbances of travel time, a stochastic fundamental diagram model will be formulated in Chap. 5; (3) by using the nonlinear stochastic signal processing method for time-frequency analysis, the typical *Eulerian* data, i.e., inductive loop detections are explored to reveal the stochastic oscillations and breakdown phenomena, then a traffic flow breakdown probability model will be established in Chap. 6.

3.2 Characteristics of Headway/Spacing/Velocity

The first question examined in this chapter is:
Question 3.1. Which one should be chosen as the fundamental modeling variables: headway, spacing, or velocity?

© Tsinghua University Press, Beijing and Springer-Verlag Berlin Heidelberg 2015
X. (M.) Chen et al., *Stochastic Evolutions of Dynamic Traffic Flow*,
DOI 10.1007/978-3-662-44572-3_3

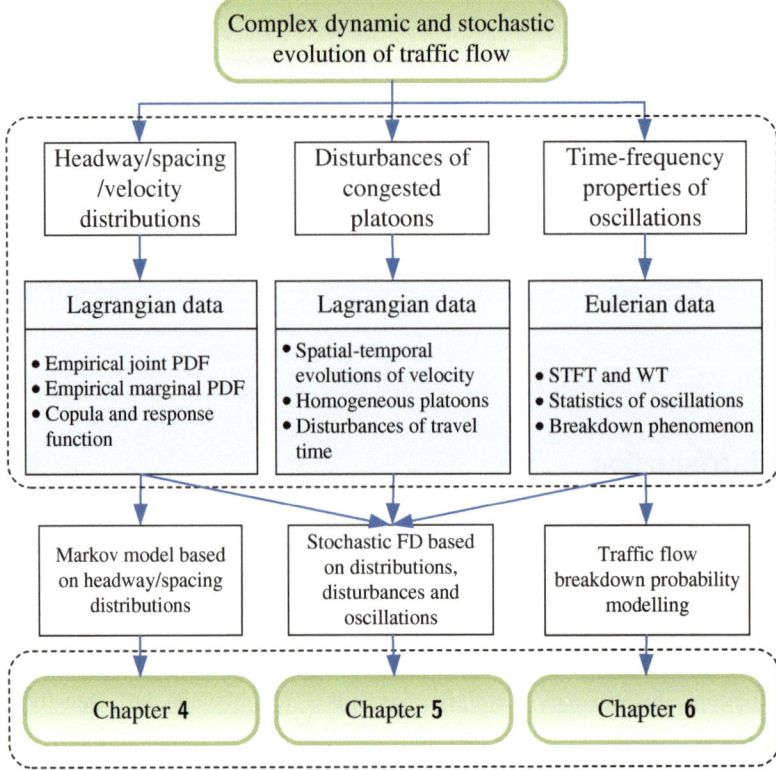

Fig. 3.1 Relationship between this chapter and other contents

To answer this question, we use different methods to retrieve and study the data from the NGSIM datasets (NGSIM 2006). Many researchers had complained about the lack of data against calibration and validation in late 1990s, e.g., Lipshtat (2009). The appearance of the NGSIM program relieves this worriment recently, since it provides high-quality and publicly available trajectory datasets that can be used to describe the microscopic driver behaviors.

NGSIM trajectory data have been emergently applied to calibrate and validate micro/macroscopic traffic models and simulations since 2005 (Chiabaut et al. 2009, 2010; Hamdar and Mahmassani 2008; Kesting and Treiber 2008; Kesting et al. 2007; Leclercq et al. 2007; Lu and Skabardonis 2007; Ossen et al. 2006; Toledo and Zohar 2007; Yeo 2008; Young and Rice 2006). Recent achievements in Intelligent Transportation System (ITS) have allowed drivers to communicate with other neighboring drivers. To identify the macroscopic properties of traffic dynamics, such as FD and shockwaves, a method to measure passing rate was proposed to determine lane-specific FD and further reinforce the linear assumption in the congested regime (Chiabaut et al. 2009). Another exploratory analysis and numerical algorithm were

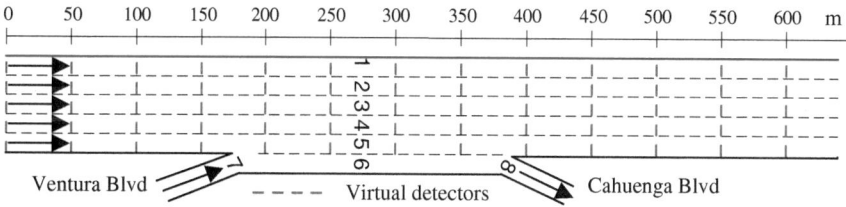

Fig. 3.2 The placements of the virtual loop detectors

applied to estimate the shockwave characteristics and propagation speeds by exploring the trajectories (Lu and Skabardonis 2007).

The trajectory data used for calibration were obtained from the southbound direction of US Highway 101, in Hollywood, Los Angeles, California. The monitoring period is from 7:50 to 8:35 am in morning rush hours on June 15, 2005. There are five through lanes (Lane 1–5) with an auxiliary (Lane 6), an on-ramp (Lane 7) and an off-ramp (Lane 8), for a total length of 640 m (see Fig. 3.2). The sampling time interval is 0.1 s. The preprocessing of the trajectory data is similar to what have been discussed in Thiemann et al. (2008) and is omitted here.

In particular, 65 virtual detectors are placed on the road as shown in Fig. 3.2. When a vehicle passes one of these virtual detectors, its current headway and spacing to its leaders will be recorded as a triple (h_i, s_i, v_i), $i \in \mathbb{N}^+$, where the footnote i is the sampling index.

Because the sampling time period is limited, we apply an aggregation method to obtain the smoother joint distributions of (h_i, s_i, v_i). More precisely, these sampled triples (h_i, s_i, v_i) are divided into 10 groups according to v_i with a uniform width of 2 m/s. The joint probability distribution of (h_i, s_i) within each range of velocities are estimated in Fig. 3.3.

Figure 3.3a gives a more clear comparison of the headway/spacing distributions with different velocity ranges. The mean values of headways decrease with the velocities and finally reach the saturation headway. Moreover, statistical tests show that these distributions approximately belong to a generic distribution family under different states (e.g., free flow and congested flow). This fact has been verified in many previous reports (e.g., Abul-Magd 2007; Knospe et al. 2002; Krbalek 2007; Krbalek and Helbing 2004; Krbalek and Šeba 2009; Krbalek et al. 2001; Li et al. 2010a; Neubert et al. 1999; Thiemann et al. 2008; Wang et al. 2009a; Zhang et al. 2007). As shown in Fig. 3.3a, there is a roughly linear correlation between headways and spacings at each velocity range, because we have the following approximate relation $s \approx vh$, where h denotes the headway, s denotes the spacing, and v is the velocity of the following vehicle.

Figure 3.3b shows that when the velocity is larger than 5 m/s, drivers will try to maintain a relatively constant headway, whose mean value is around 2 s and the distributions belong to a family of lognormal distributions. On the other hand, Fig. 3.3c shows that when the velocity is lower than 5 m/s, the drivers will try to maintain a relatively constant spacing, whose mean value is around 8 m and the distributions

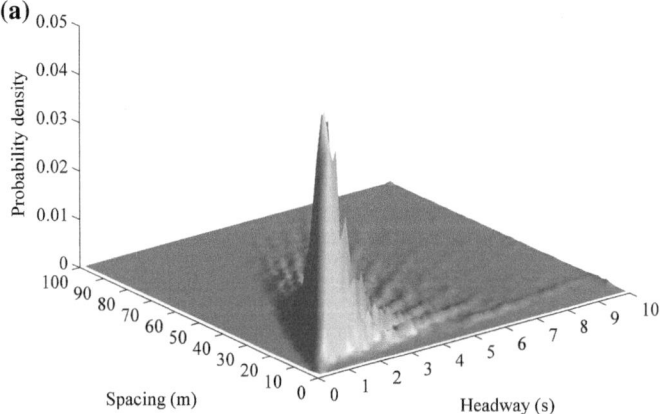

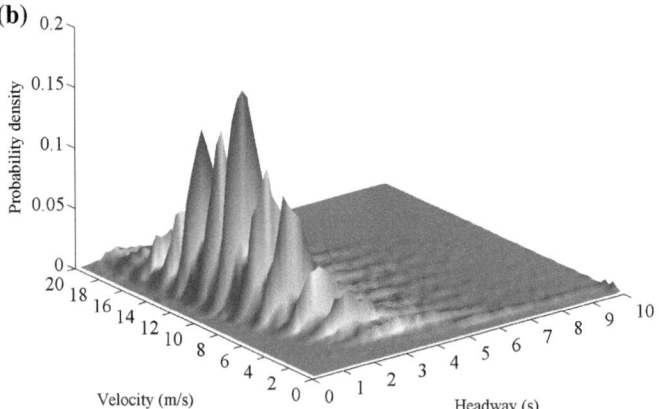

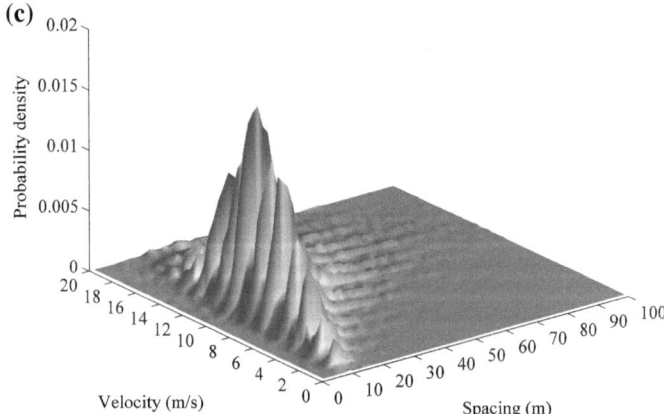

Fig. 3.3 The empirical joint probability distributions of headway/spacing/velocity. **a** (h_i, s_i). **b** (h_i, v_i). **c** (s_i, v_i)

also belongs to a certain lognormal family of distribution. Figure 3.3c further shows that the mean values of spacings increase with the velocities and gradually reach the saturation spacing.

Since headway/spacing/velocity are correlated to each other, it is significant to study their stochastic characteristics for deeply understanding traffic flow evolutions. In probability theory and statistics, a copula can be used to describe the dependence between random variables by estimating marginal distributions. There are many parametric copula families available to control the strength of dependence.

Take the bivariate random variables (Z_1, Z_2) [1] as an example, define the joint CDF as

$$F(z_1, z_2) \triangleq \Pr\{Z_1 \leqslant z_1, Z_2 \leqslant z_2\} \tag{3.1}$$

According to Sklar's theorem (Nelsen 2006), given the joint CDF and marginal CDF of bivariate random variables, the copula is unique, i.e.,

$$F(z_1, z_2) = \mathcal{C}(F_1(z_1), F_2(z_2)) \tag{3.2}$$

Incorporate random variables ξ, ζ defined in $[0, 1]$, the copula is

$$\mathcal{C}(\xi, \zeta) = F(F_1^{-1}(\xi), F_2^{-1}(\zeta)) \tag{3.3}$$

where $\xi = F_1(z_1), \zeta = F_2(z_2), z_1 = F_1^{-1}(\xi), z_2 = F_2^{-1}(\zeta)$. According to Frèchet–Hoeffding theorem (Nelsen 2006), the copula bounds are

$$\max\{\xi + \zeta - 1, 0\} \leqslant \mathcal{C}(\xi, \zeta) \leqslant \min\{\xi, \zeta\} \tag{3.4}$$

Suppose we have n observations $(Z_1^{(i)}, Z_2^{(i)})$, $i \in \mathbb{N}$, where $\mathbb{N} = \{1, 2, \ldots, n\}$, the corresponding true copula observations would be

$$(U_1^{(i)}, U_2^{(i)}) = (F_1(Z_1^{(i)}), F_2(Z_2^{(i)})), \quad i \in \mathbb{N} \tag{3.5}$$

If the marginal distributions are unknown, the pseudo copula observations can be constructed by using the empirical distributions

$$\hat{F}_1(z_1) = \frac{1}{n} \sum_{i=1}^{n} \mathbf{1}(Z_1^{(i)} \leqslant z_1), \quad \hat{F}_2(z_2) = \frac{1}{n} \sum_{i=1}^{n} \mathbf{1}(Z_2^{(i)} \leqslant z_2) \tag{3.6}$$

where $\mathbf{1}(\cdot) \mapsto \{0, 1\}$ is indicator function.

The pseudo copula observations are defined as

---

[1] (Z_1, Z_2) represent bivariate normal random variables (h, v) or (s, v). Since (h, s) may not follow the bivariate lognormal distributions, (h, s) are not considered.

$$(\hat{U}_1^{(i)}, \ \hat{U}_2^{(i)}) = (\hat{F}_1(Z_1^{(i)}), \ \hat{F}_2(Z_2^{(i)})), \quad i \in \mathbb{N} \tag{3.7}$$

The corresponding empirical copula is then defined as

$$\hat{C}(\xi, \ \zeta) = \frac{1}{n} \sum_{i=1}^{n} \mathbf{1}(\hat{U}_1^{(i)} \leqslant \xi, \ \hat{U}_2^{(i)} \leqslant \zeta) \tag{3.8}$$

On the contrary, if given the marginal distributions of $(F_1(z_1), \ F_2(z_2))$ and the copula $C(\xi, \ \zeta)$, Monte-Carlo integral can be used to estimate the expectation of a response function $\{\mathcal{R}(z_1, \ z_2) : \mathbb{R}^2 \mapsto \mathbb{R}\}$,

$$
\begin{aligned}
\mathrm{E}[\mathcal{R}(z_1, \ z_2)] &= \int_{\mathbb{R}^2} \mathcal{R}(z_1, \ z_2) \, \mathrm{d}F(z_1, \ z_2) \\
&= \int_{[0, \ 1]^2} \mathcal{R}(F_1^{-1}(\xi), \ F_2^{-1}(\zeta)) \, \mathrm{d}C(\xi, \ \zeta) \\
&= \int_0^1 \int_0^1 \mathcal{R}(F_1^{-1}(\xi), \ F_2^{-1}(\zeta)) C'(\xi, \ \zeta) \, \mathrm{d}\xi \, \mathrm{d}\zeta
\end{aligned}
\tag{3.9}
$$

where $C'(\xi, \ \zeta) = \partial^2 C(\xi, \ \zeta)/(\partial \xi \partial \zeta)$ is the density function of copula.

First, generate n i.i.d. samples based on copula C, i.e.,

$$(U_1^{(i)}, \ U_2^{(i)}) \sim C(\xi, \ \zeta), \ i \in \mathbb{N} \tag{3.10}$$

then, use the inverse marginal CDF $(F_1^{-1}, \ F_2^{-1})$ to produce samples

$$(Z_1^{(i)}, \ Z_2^{(i)}) = (F_1^{-1}(U_1^{(i)}), \ F_2^{-1}(U_2^{(i)})), \ i \in \mathbb{N} \tag{3.11}$$

then, calculate the approximate value of the response function by its empirical value

$$\mathrm{E}[\mathcal{R}(z_1, \ z_2)] \approx \frac{1}{n} \sum_{i=1}^{n} \mathcal{R}(Z_1^{(i)}, \ Z_2^{(i)}) \tag{3.12}$$

Specifically, assume that random variables Z_1 and Z_2 follow lognormal distributions, their PDFs are

$$f(z_1) = \text{Log-}\mathcal{N}(z_1; \ \mu_1, \ \sigma_1) = \frac{1}{\sqrt{2\pi}\sigma_1 z_1} \exp\left(-\frac{(\log z_1 - \mu_1)^2}{2\sigma_1^2}\right), \quad z_1 > 0 \tag{3.13a}$$

$$f(z_2) = \text{Log-}\mathcal{N}(z_2; \mu_2, \sigma_2) = \frac{1}{\sqrt{2\pi}\sigma_2 z_2} \exp\left(-\frac{(\log z_2 - \mu_2)^2}{2\sigma_2^2}\right), \quad z_2 > 0$$

(3.13b)

where μ_1, σ_1, μ_2, σ_2 are the parameters, satisfying $\log Z_1 \sim \mathcal{N}(\mu_1, \sigma_1^2)$, $\log Z_2 \sim \mathcal{N}(\mu_2, \sigma_2^2)$.

The joint PDF of Z_1 and Z_2 is

$$f(z_1, z_2) = \frac{1}{2\pi\sqrt{1 - \rho_z^2}\sigma_1\sigma_2 z_1 z_2} \exp\left[-\frac{1}{1 - \rho_z^2}\left(\frac{(\log z_1 - \mu_1)^2}{2\sigma_1^2}\right.\right.$$
$$\left.\left. -\frac{\rho_z(\log z_1 - \mu_1)(\log z_2 - \mu_2)}{\sigma_1\sigma_2} + \frac{(\log z_2 - \mu_2)^2}{2\sigma_2^2}\right)\right]$$

(3.14)

where ρ_z is the correlation coefficient of Z_1 and Z_2, i.e.,

$$\rho_z = \frac{\mathrm{E}[(\log z_1 - \mu_1)(\log z_2 - \mu_2)]}{\sigma_1\sigma_2}, \quad |\rho_z| \leqslant 1$$

(3.15)

The conditional PDFs of Z_1 and Z_2 are

$$f(z_1|z_2) = \frac{f(z_1, z_2)}{f(z_2)} = \frac{1}{\sqrt{2\pi}\sigma_{z_1|z_2} z_1} \exp\left(-\frac{(\log z_1 - \mu_{z_1|z_2})^2}{2\sigma_{z_1|z_2}^2}\right)$$

(3.16a)

$$f(z_2|z_1) = \frac{f(z_1, z_2)}{f(z_1)} = \frac{1}{\sqrt{2\pi}\sigma_{z_2|z_1} z_2} \exp\left(-\frac{(\log z_2 - \mu_{z_2|z_1})^2}{2\sigma_{z_2|z_1}^2}\right)$$

(3.16b)

It can be found that $f(z_1|z_2)$ and $f(z_2|z_1)$ are still lognormal type, but the parameters are

$$\mu_{z_1|z_2} = \mu_1 + \rho_z\frac{\sigma_1}{\sigma_2}(\log z_2 - \mu_2), \quad \sigma_{z_1|z_2} = \sqrt{1 - \rho_z^2}\sigma_1$$

(3.17a)

$$\mu_{z_2|z_1} = \mu_2 + \rho_z\frac{\sigma_2}{\sigma_1}(\log z_1 - \mu_1), \quad \sigma_{z_2|z_1} = \sqrt{1 - \rho_z^2}\sigma_2$$

(3.17b)

3.3 Congested Platoon Oscillations

The efforts to capture traffic oscillations can be traced back to early studies on temporal and asymptotic stabilities of linear car-following models investigated with frequency analysis tools (Chandler et al. 1958; Gazis et al. 1961). Later on, Gipps Gipps (1981) proposed a nonlinear car-following model to better reproduce traffic evolution. Bando et al. (1995) developed the well-known OVM model to study

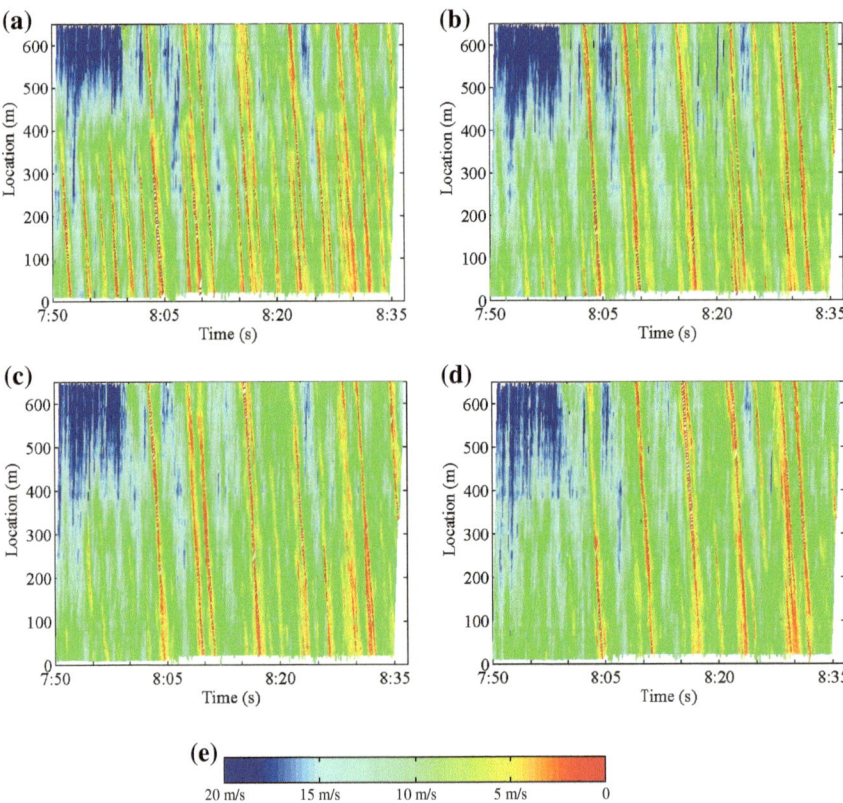

Fig. 3.4 Spatial-temporal evolution of velocity. **a** Lane 1. **b** Lane 2. **3** Lane 3. **d** Lane 4. **e** Speed yardstick

stop-and-go traffic. Numerical traffic approaches were usually incorporated to demonstrate the oscillation phenomena (Bertini and Leal 2005; Cassidy and Bertini 1999; Helbing et al. 1999; Kerner and Rehborn 1996a, b; Kühne 1987).

Li et al. (2010b) mentioned that *traffic oscillations, also known as the "stop-and-go" traffic, refer to the phenomenon that congested traffic tends to oscillate between slow-moving and fast-moving states rather than maintain a steady state. Traffic oscillations create significant driving discomfort, travel delay, extra fuel consumption, increased air pollution, and potential safety hazards.*

Figure 3.4 shows the spatial-temporal evolution of velocity by using trajectories on Lane 1 through Lane 4. During the first 15 min, the traffic is almost stable between 400 m (near an off-ramp at 398 m) and the segment end, but the traffic state shows several small oscillations propagating upstream with a relatively constant speed. Particularly on Lane 1, the oscillation period is much smaller than the other three lanes. During the following 30 min, the stop-and-go phenomenon occurs from the segment end to the most upstream location. On Lane 1, Lane 2, and Lane 3, these

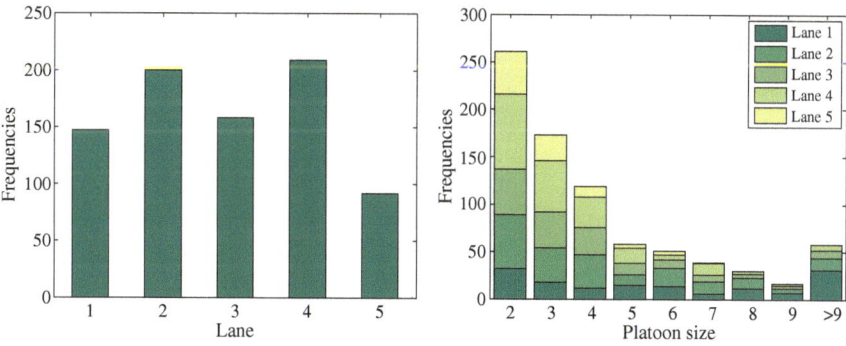

Fig. 3.5 Statistics of the stable platoons by lanes and size of platoons

shockwaves are mixed with other oscillations caused by the off-ramp, while some shockwaves vanish on Lane 4. The velocity evolutions clearly show the generation, propagation and dissipation of stop-and-go waves, as well as the empirical oscillation period and intensity.

We extracted vehicles in a stable platoon (not fewer than two vehicles) that means each vehicle kept in the same lane without lane changing between $x = 50$ m and $x = 600$ m and the spacing between two consecutive vehicles were smaller than 50 m. As shown in Fig. 3.5, there were 806 stable platoons extracted, with 147, 200, 158, 209, and 92 stable platoons on each lane, respectively. There were 3,545 full trajectories, or 45.84 % of the total 7,734 measured vehicles. The number of vehicles in one stable platoon decreased with the size of platoons, which indicated that a platoon with a larger size was more likely disrupted. We will discuss the properties of travel time distribution based on these stable platoons in the following sections.

In Fig. 3.6, all of the four lanes showed the same feature of oscillatory travel times. The influences of traffic jam shockwaves on travel times are obviously shown in Fig. 3.6. The minimal travel time is about 40 s. However, encountering shockwaves will increase travel time to 80–100 s. Obviously, the empirical distributions of such travel times are multimodal, and cannot be simply depicted by a single lognormal distribution.

In order to understand the traffic flow dynamic in the NGSIM Highway 101 dataset, we plot several trajectories typical of platoons shown in Fig. 3.7. The first platoon shown in Fig. 3.7a moves at a homogeneous speed (close to the free-flow speed) and does not encounter any shockwaves. The corresponding average travel time and variance are $\bar{T} = 41.1$ s, $\sqrt{S^2[T]} = 0.8$ s, respectively. Figure 3.7b–d plot three other platoons interrupted by one, two, and multiple shockwaves, respectively. The number and durations of jams induce oscillatory travel time that consists of different number of components. It is worthy to point out that the case of multiple shockwaves is rarely observed in the explored NGSIM dataset, while the vehicles shown in Fig. 3.7a–c are in the majority.

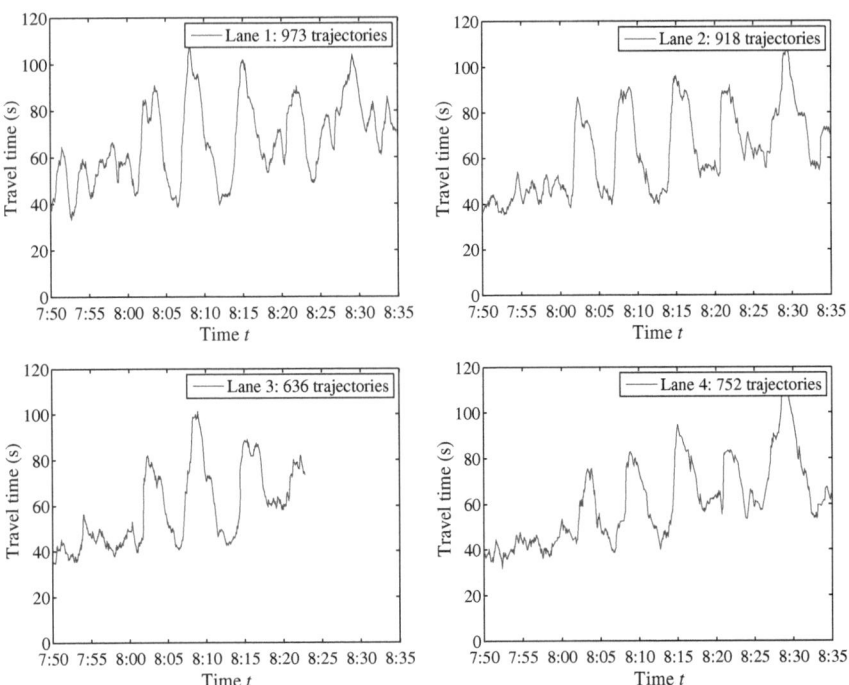

Fig. 3.6 Time series of travel times for NGSIM Highway 101 dataset

3.4 Time-Frequency Properties

Traffic congestion pattern performs an important role in traffic flow analysis. Signal processing is a powerful tool for the analysis and synthesis of time series. Characteristics localization of time series in spatial (or time) and frequency (or scale) domains can be accomplished to learn the time-frequency properties of empirical traffic flow.

Let $\{z_t, \ t = 0, 1, \ldots, T - 1\}$ be a discrete series of *Eulerian* data of traffic flow, mean speed or time occupancy. Then the discrete Fourier transform (DFT) is

$$\hat{z}_d(\omega) = \sum_{t=0}^{T-1} z_t e^{-i\omega t}, \quad \forall \omega \in [0, \ \pi] \tag{3.18}$$

where ω is the digital frequency that can be transformed as

$$\hat{z}(\Omega) = \hat{z}_d(\Omega \Delta) = \sum_{t=0}^{T-1} z_t e^{-i\Delta\Omega t}, \quad \forall \Omega \in [0, \ \pi/\Delta] \tag{3.19}$$

where $\Omega = \omega/\Delta$ is analog frequency, Δ is the period of field data collection.

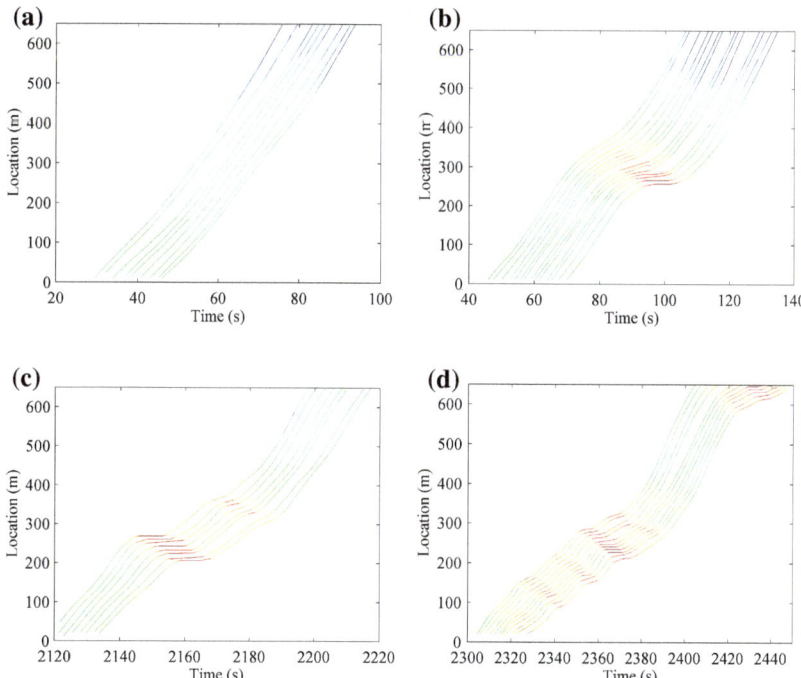

Fig. 3.7 Oscillations of typical platoon trajectories. **a** $\bar{T} = 41.1$ s, $\sqrt{S^2[T]} = 0.8$. **b** $\bar{T} = 56.0$ s, $\sqrt{S^2[T]} = 2.1$. **c** $\bar{T} = 73.9$ s, $\sqrt{S^2[T]} = 1.3$. **d** $\bar{T} = 93.7$ s, $\sqrt{S^2[T]} = 1.6$

The discrete form of short-time Fourier transform (STFT) is

$$\text{STFT}(k, \ \Omega) = \sum_{t=0}^{T-1} z_t w(t-k) e^{-i\Delta\Omega t}, \quad \forall \Omega \in [0, \ \pi/\Delta] \tag{3.20}$$

where k is the time of measurements (mean speed, density, etc.), $w(t-k)$ is the window function, a rectangular window is chosen for analysis, i.e., $w(t-k| - l_w\Delta/2 \leqslant t-k \leqslant l_w\Delta/2) = 1$, otherwise 0, $l_w\Delta$ is the window length.

Apply the STFT approach to empirical traffic flow data collected on a 8.8-kilometer segment of G6 Highway (Beijing-Tibet Expressway or China National Expressway 6) in Beijing, with 17 loop detectors, named "51001–51017". The measurement time interval is $\Delta = 120$ s. As an example, Fig. 3.8 shows the application of the STFT approach on mean speed of September 23, 2008 (Tuesday). The horizontal axis shows the analogy frequency Ω. The warmer color represents more obvious oscillations, while the colder color represents free-flow states. It can be seen that the STFT approach clearly exhibits the time-frequency characteristics of traffic speed

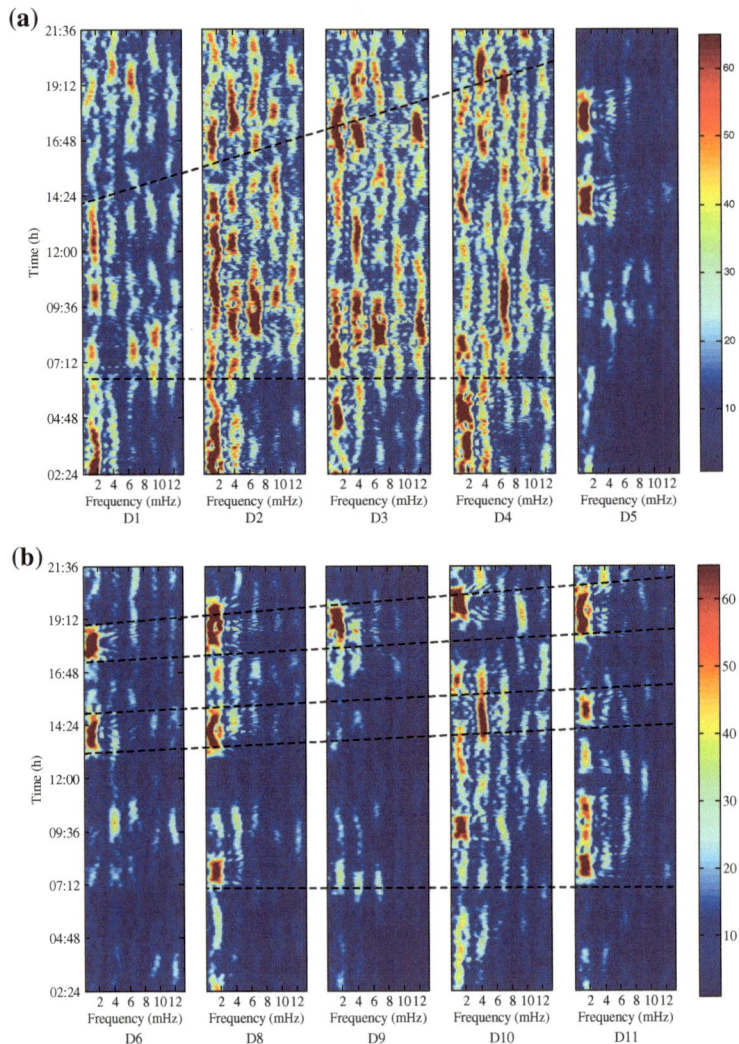

Fig. 3.8 Illustration of the STFT approach with empirical loop data. **a** Loop detector D1~D2. **b** Loop detector D6, D8~D11

data along a series of loop detectors, and describes the stochastic oscillations in an illustrative way. The low-frequency oscillations with a longer cycle need more time to dissipate, but the high-frequency oscillations have significant influences on traffic stability. Detectors D2~D4 are in relatively congested states with obvious stochastic oscillations, especially between 7:00 and 19:00. Meanwhile, the local extreme values of detectors D2~D4 are approximate 2,[2] 4 and 6 mHz, and the corresponding

----

[2] 1 mHz = 0.001 Hz.

oscillating circles are 8.3, 4.2, 2.8 min, respectively. For detector D5, the duration of traffic congestion is from 14:24 to 18:00 or so. The principle frequency is about 1 mHz and the corresponding oscillating circle is 16.7 min.

Figure 3.8b shows that the stochastic oscillations of detectors D6, D8~D11 concentrate in three peaks, i.e., morning rush hours (7:00–9:00), afternoon rush hours (14:00–15:00), and evening rush hours (18:00–19:30).

More precisely, wavelet transform (WT) provides an interesting and useful alternative to the classical STFT. The basic difference consists in the fact that WT performs a multiresolution analysis of the signals, by making use of short windows at high frequencies and long windows at low frequencies. This property will be more appropriate for the applications to stochastic and oscillatory traffic flow data.

Continuous wavelet transform (CWT) possesses the ability to construct a time-frequency representation of a signal that offers a good time and frequency localization. The mother wavelet $\psi(t)$ is a continuous function in both the time and frequency domains, satisfying

$$\int_{-\infty}^{\infty} |\psi(t)|^2 \, dt = \int_{-\infty}^{\infty} |\hat{\psi}(f)|^2 \, df < \infty, \quad \int_{-\infty}^{\infty} \psi(t) \, dt = 0 \qquad (3.21)$$

where $\hat{\psi}(f) = \int_{-\infty}^{\infty} \psi(t)e^{-i(2\pi f)t} \, dt$ is the FT of $\psi(t)$.

The most common mother wavelet is Gauss wavelet

$$\psi(t) = \sqrt{2\pi}\phi(t) = e^{-t^2/2} \qquad (3.22)$$

Its first order derivative is

$$\psi(t) = \sqrt{2\pi}\phi'(t) = -te^{-t^2/2} \qquad (3.23)$$

Its second order derivative is as a Mexican hat wavelet

$$\psi(t) = -\sqrt{2\pi}\phi''(t) = (1 - t^2)e^{-t^2/2} \qquad (3.24)$$

We can include the scale parameter a and translation parameter b, i.e.,

$$\psi\left(\frac{t-b}{a}\right) = \left[1 - \left(\frac{t-b}{a}\right)^2\right]\exp\left[-\frac{1}{2}\left(\frac{t-b}{a}\right)^2\right] \qquad (3.25)$$

Based on the wavelets, CWT coefficients can be obtained by

$$\text{CWT}(a, b) = \frac{1}{\sqrt{a}}\int_{-\infty}^{\infty} z(t)\psi^*\left(\frac{t-b}{a}\right) dt = \frac{1}{\sqrt{a}}\int_{-\infty}^{\infty} \hat{z}(f)\hat{\psi}^*(af)e^{i(2\pi f)b} \, df \qquad (3.26)$$

where CWT(a, b) is the wavelet transform at time b given the scale a.

Then, the mean energy at time b across all scales is formulated as

$$E(b) = \frac{1}{\max(a)} \int_0^{\max(a)} |\text{CWT}(a, b)|^2 \, da \qquad (3.27)$$

In real applications, Fig. 3.9 incorporates CWT to identify the begin and end points of stochastic oscillations. Take the mean speed measured at detector D8 as an example, Fig. 3.9a shows how to apply the most simplest Haar wavelet to field data. We can find the energy evolution with time. The resolution and accuracy decrease with the scale. So the characteristic traffic state transition points can be approximately identified.[3]

As shown in Fig. 3.9b, the CWT coefficients become zero at 13:30 and 17:20 for more scales when the second order Gauss wavelet is used. In this way, we can analyze the concave-convex properties of the curved surface of CWT coefficients. The inflection points can be identified in such a way. In Fig. 3.9b, we can find the begin and end points are 13:30–14:20 and 17:20–18:00, respectively.

For further applications of the WT approach, we study the traffic breakdown phenomenon that can be defined as a sudden state transition from free-flow state to congested state. It usually occurs when the mean speed drops rapidly below a certain threshold in a short period of time Banks (2006). From the viewpoint of detection theory in statistical signal processing, traffic breakdown can be inferred as the problem of model change detection. In this study, we divide the whole traffic breakdown process into breakdown transition regime, congested regime, and breakdown recovery regime. Empirical measurements of highway traffic breakdown using discrete wavelet transform (DWT) reveal the breakdown activation and recovery process of transitions from free-flow to congested states.

Figure 3.10a shows that a 9 km segment of I80-W (westbound) in and near Berkeley, California. It contains four to five lanes and 20 detector stations named D1 thought D20. The southbound traffic direction is toward the downtown area of San Francisco through the Bay Bridge. The data used in this study was downloaded via the Performance Measurement System (PeMS 2011), which is jointly developed by California Department of Transportation (Caltrans), University of California at Berkeley, and California PATH (Partnership for Advanced Technology on Highways). Each detector measures the raw, 30 s data of vehicle counts and time occupancies that are then transformed into the PeMS schema and archived in the PeMS database. The 30 s data are imputed and aggregated to 5 min samples per lane without holes. The median lane is a high-occupancy vehicle (HOV) lane with different characteristics from other regular lanes, this study only considers the three shadowed lanes (see Fig. 3.10a). The measured 5 min interval and lane-aggregated data from October 1 to 31, 2011 (31 days). The maximal and minimal separations of loop detectors are 1.0

---

[3] The characteristic points can be more accurately identified by using other more complex wavelets.

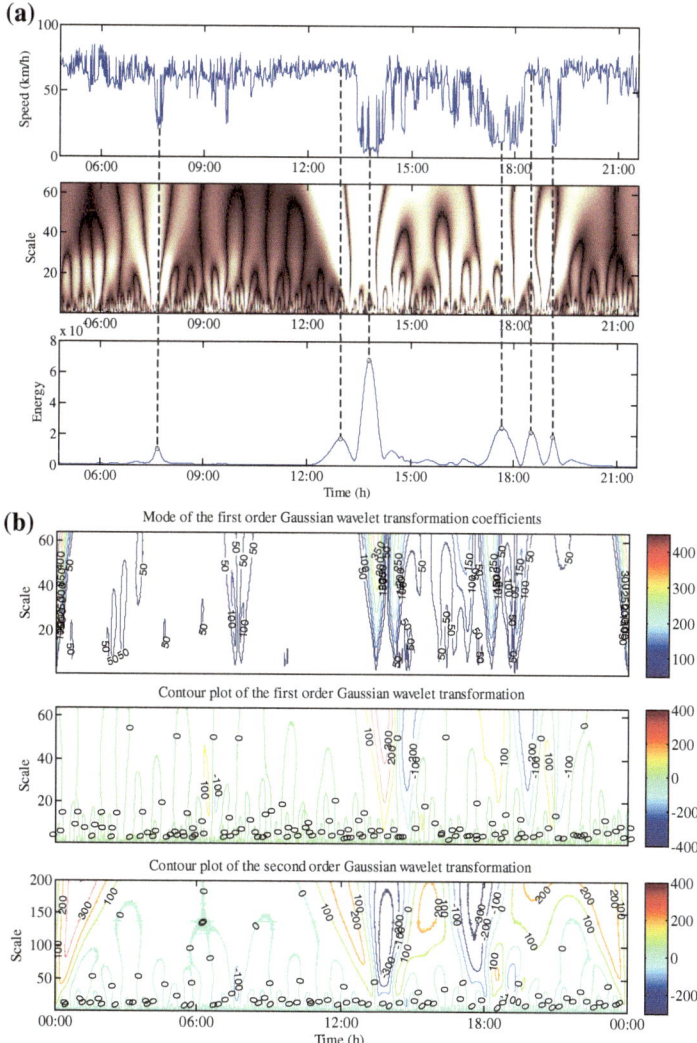

Fig. 3.9 Time-frequency features of traffic flow by wavelet transform. **a** Time-frequency features of mean speed (D8). **b** Contour plot of the first and second order Gaussian WT coefficients (D8)

and 0.06 km, respectively, and the mean separation is 0.48 km, which is close enough for empirically observing spatial-temporal evolutions.

Figure 3.10a shows a segment of Highway G6 in Beijing. It is a 25 km segment equipped with 34 double inductive loop detectors (labeled here as D1 through D34) on each lane in both northbound and southbound directions. The maximal and minimal separations of loop detectors are 3.0 and 0.1 km, respectively, and the average value is 0.75 km, which is close enough as well although it is a little larger than the I80-W

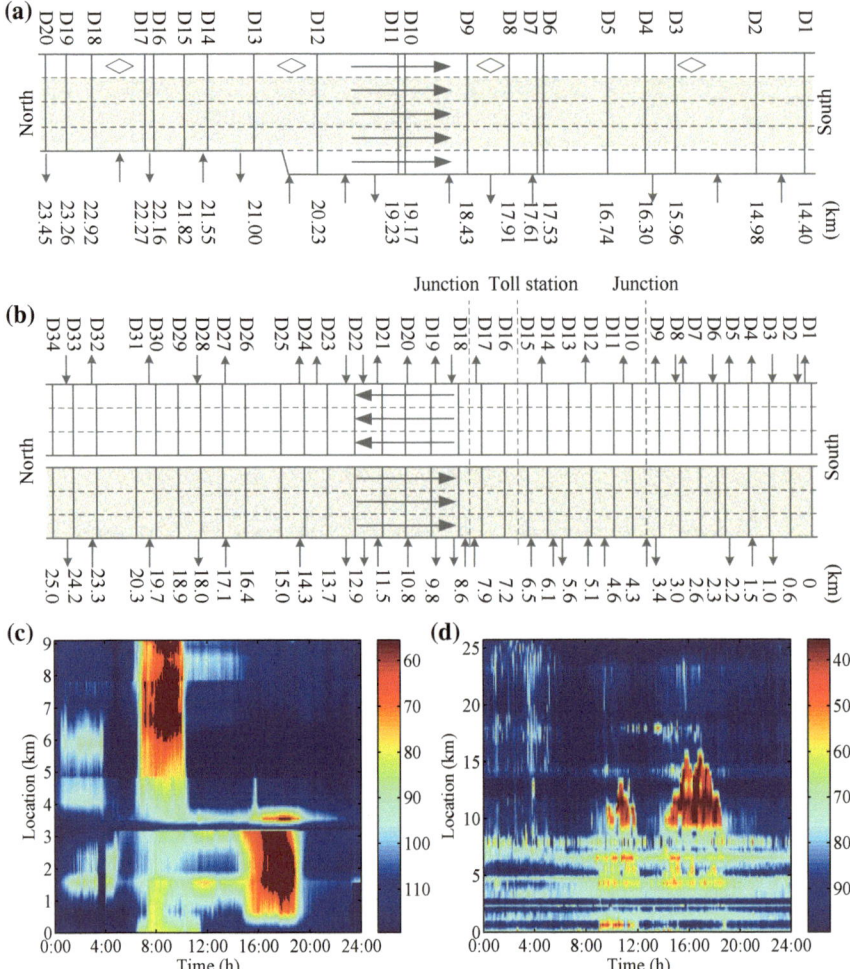

Fig. 3.10 Schematic of loop detectors and spatial-temporal evolutions of traffic speed. **a** Schematic of Highway I80-W. **b** Schematic of Highway G6. **c** Spatial-temporal evolutions on I80-W. **d** Spatial-temporal evolutions on G6-S

site. Vehicle counts (volumes for automobiles and trucks) and time occupancies were collected by loop detectors on each lane and aggregated for every 5 min. The time-average speed at each loop detector was not measured directly, but estimated based on the vehicle counts, the occupancies, the average vehicle length, and the effective detection zone length. The collected 5 min interval and lane-aggregated data are from October 1 to 31, 2010 (31 days). During the measurement period, the experimental segment experienced two severe traffic accidents on October 1 and 31, respectively, but nonrecurrent traffic congestions on these two days are out of the scope of this book. It indicated that the vehicle conservation was roughly maintained by comparing the cumulative counts from upstream and downstream locations although there were many on/off-ramps without loop detectors installed in both bounds. Additionally, there are two main junctions (between D9 and D10, between D17 and D18) and one toll station (between D15 and D16) along the highway. It is worthy to point out that the toll station is a physical bottleneck and incurs congestions. In order to make the selected site more comparable with the I80-W segment, we choose the southbound direction toward to the downtown of Beijing in this study (see the three shadowed lanes in Fig. 3.10a).

Figure 3.10c shows the mean speeds and time occupancies across all lanes for each 5 min interval in the space-time plot. The warmer shades indicate lower speeds and higher time occupancies, contrarily the cooler shades represent higher speeds and lower time occupancies. In the southbound traffic flow evolutions on October 12, 2011 (Wednesday), there are two physical bottlenecks along this segment: (1) the merging area with another Interstate Highway I580-E at the upstream of the D12 (20.23 km); (2) the Bay Bridge toll station at the downstream of the D1 (14.40 km). In typical weekdays, recurrent congestions usually occur at the first bottleneck and propagate upstream rapidly in morning rush hours, and emerge at the second bottleneck in evening rush hours.

Figure 3.10d shows the recurrent traffic congestions on October 5, 2010 (Tuesday). Based on the empirical analysis, there is a physical bottleneck of toll station along this segment between D15 (6.5 km) and D16 (7.2 km). Generally, the bottleneck induces congestions in morning rush hours and more severely in the evening rush hours. Obviously different from the case of I80-W, it shows that the toll station incurred congestions in weekends (e.g., October 3, 2010, Tuesday) even heavier than weekdays, this was even worse due to a higher demand of returning traffic back to the city on Sunday evening rush hours and much more trucks run on G6-S without entry limits in weekends.

In field applications, it is more efficient and effective for computation to reconstruct the original signal by using infinite summations of discrete wavelet coefficients rather than continuous integrals. As shown in Fig. 3.11, we select the biorthogonal spline wavelets as an example to illustrate the application of DWT in traffic breakdown identifications, because biorthogonal spline wavelets enable symmetric and antisymmetric wavelets and allow certain desirable properties to be incorporated separately within the decomposition wavelets and the reconstruction wavelets.

Figure 3.11a, b illustrate two examples of speed time series, collected at D13, October 12, 2011 and D6, October 13, 2011, respectively. For the sake of simplicity, we demonstrate that equal to or larger than 90 km/h represents the free-flow state, between 50 and 90 km/h indicates the less congested state or "synchronized flow" in three-phase theory, while smaller than 50 km/h denotes the congested state. The average wavelet-based energy across all scales at a specific time tracks the abrupt speed change well. It can be seen from these two examples that DWT identifies some other speed fluctuations within the same traffic states as well.

Here, we only take into account the temporally distributed energy peaks that correspond to traffic state transitions in speed time series, particularly for the peaks

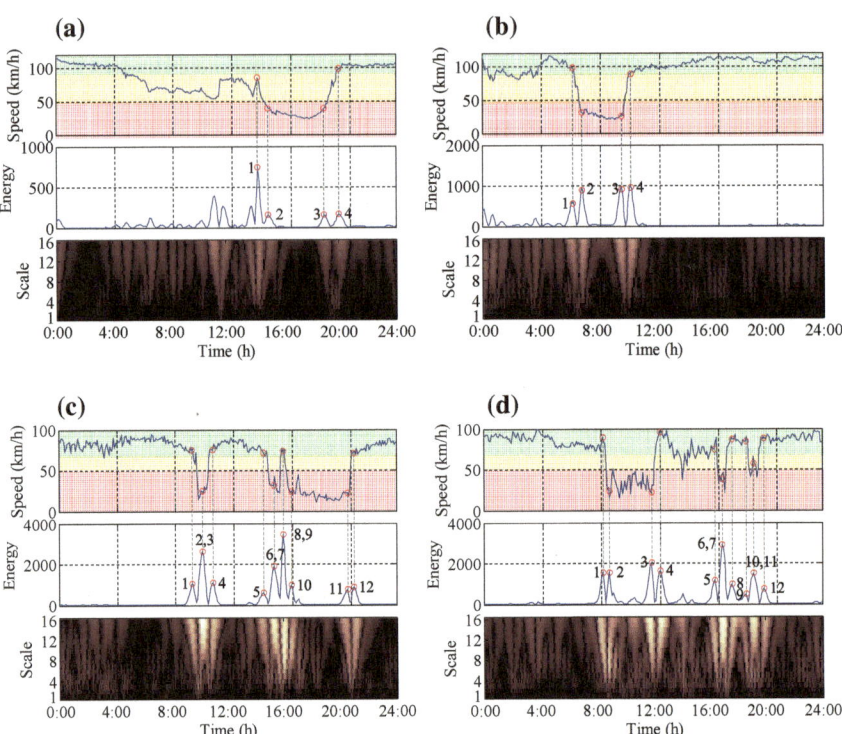

Fig. 3.11 Identifications of traffic breakdown occurrence and recovery time. **a** I80-W, D13, October 12, 2011. **b** I80-W, D16, October 13, 2011. **c** G6-S, D20, October 3, 2010. **d** G6-S, D2, October 5, 2010

where traffic flow changes from free-flow state to congested state. The characteristics of traffic breakdown can be approximately identified by four points: Point 1 is the breakdown occurrence time T_1 before traffic state transition; Point 2 is the end of transition from free-flow state to less congested state or congested flow, i.e., T_2; Point 3 represents the beginning of breakdown recovery time T_3; and Point 4 denotes the end of breakdown, i.e., recovery time T_4. We also show the absolute values of discrete wavelet transform coefficients from scales 1–16 in the examples.

Figure 3.11c, d demonstrate the applications of DWT in detecting breakdown characteristic points via two speed time series, collected at D20, October 3, 2010 and D2, October 5, 2010. Due to the different speed limit of G6-S, we define the three traffic states as follows, equal to or larger than 70 km/h represents free-flow state, between 50 and 70 km/h indicates the less congested state, while smaller than 50 km/h denotes the congested state. It can be seen that speed profiles of G6-S involve more oscillations and much more complicated features of traffic breakdown. Analogically, the characteristic points of traffic breakdown can be approximately identified by Points 1–4, these two examples show multiple breakdown phenomena occurred and we defined the characteristic points sequentially, e.g., we had 12 points in these two profiles and we treated Point 5 as Point 1, Point 6 as Point 2 and so on. Besides, it was found that when traffic breakdown lasted for a short while, the wavelet-based approach was not capable of identifying the difference between T_2 and T_3, which was a shortcoming of this method but could be improved by selecting higher resolution wavelet functions.

Based on the previous simplification of traffic breakdown phenomena, we continue to define some statistics to examine the statistical characteristics of breakdown. $T_2 - T_1$ quantifies the process of breakdown transition, $T_4 - T_3$ expresses the process of breakdown recovery, $v(T_1) - v(T_2)$ identifies the value of speed decline before and after breakdown, $v(T_4) - v(T_3)$ shows the value of speed increase during recovery from traffic breakdown. Furthermore, we can also define $T_4 - T_1$ as the whole duration of traffic breakdown. Due to the nonstationarity and complex features of traffic flow in congestion, it is difficult to mathematically formulate the exact evolution of traffic states. Instead, we empirically observed the datasets from PeMS (2011) and collected the characteristic peaks that satisfied the breakdown transition and recovery conditions via the wavelet-based identification approach.

Based on the 31-day data (from October 1 to 31, 2011) from 20 loop detecting stations on I80-W, Fig. 3.12a, b show the statistical results of $T_2 - T_1$ and $T_4 - T_3$, based on more than 436 valid samples. The mean values of breakdown transition and recovery duration are 47.8 and 41.7 min, respectively.

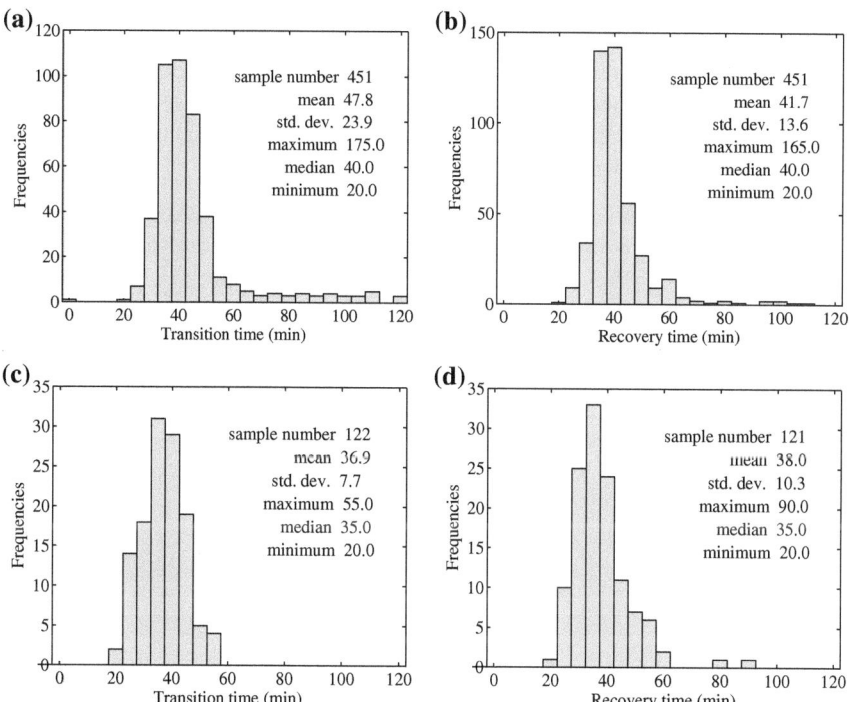

Fig. 3.12 Statistical characteristics of traffic breakdown transition time and recovery time. **a** I80-W. **b** I80-W. **c** G6-S. **d** G6-S

Based on the 7-day data (from October 2 to 8, 2010) from 34 loop detecting stations on G6-S, more than 117 valid samples, Fig. 3.12c, d show that mean values of breakdown transition and recovery time are 36.9 and 38.0 min, that are about 22 and 9 % shorter than the case of I80-W with 47.8 and 41.7 min, respectively. We can see that the G6-S case shows a much steeper feature of traffic breakdown and performs a little shorter recovery time from breakdown.

The different performances of traffic breakdown transition time, recovery time, and speed changes between the two study sites in the US and China can be interpreted as follows:

1. Different patterns of driving behaviors may induce the distinct differences of breakdown transition time and recovery time. For example, the drivers in China are more heterogeneous from the perspective of driving maneuvers, and further generate more aggressive car-following behaviors, shorter space headways, or more jerks behind traffic queues, which encourages the abrupt breakdown transition or recovery. Another appropriate reason closely related to the breakdown process is regarding to the lane-changing or merging behaviors that perform more frequently and unexpectedly in China, e.g., merging vehicles on shoulder lane usually cause oscillations that propagate across to the median lane much more quickly than in the US. Overall, the diversity of driving behaviors influences traffic breakdown to a large extent.
2. Different flow-density-speed relations or fundamental diagrams along the two segments caused by different speed limits, capacities, and varieties of vehicles may influence the distributions of speed decline and increase during the whole traffic breakdown duration. The speed limit and highway capacities in China are generally smaller than in the US, which partly influences the congested speed at the end of breakdown transition.

The foregoing approach effectively distinguishes useful breakdown information from nonlinear and nonstationary speed time series. Comparatively, the analysis of loop detector data from two highways (I80-W in Berkeley, CA, U.S. and G6-S in Beijing, China) shows that the wavelet-based energy approach enables effective identification of traffic breakdown occurrence time, recovery time, state transition time, and congested speed. Observations show that I80-W preforms longer transition time and recovery time in traffic breakdown than G6-S, from which it infers that traffic breakdown occurs and recovers more abruptly on G6-S. The empirical study has important implications for understanding the breakdown process and statistical characteristics of traffic state transitions.

3.5 Summary

Transportation system is a complex, dynamic, stochastic system. Herrera and Bayen (2010) classified two kinds of traffic flow measurements depending on the way of data collection. The *Eulerian* measurements consist of observing traffic flow at fixed

locations such as loop detecting stations, while the *Lagrangian* measurements identify specific vehicle trajectories by using GPS, location-based technology inside
cellular phones, vehicle tracking device, etc.

This chapter makes use of two typical kinds of traffic flow datasets for the qualitative to quantitative analyses of traffic flow evolution characteristics. Specifically, the
loop detectors on I80-W in Berkeley and G6 Highway in Beijing are chosen as the
Eulerian measurements, including the flow rate, mean speed, time occupancy, and
large vehicle percentage for each separate lane. The NGSIM trajectory datasets are
utilized as the *Lagrangian* measurements, including location, velocity, acceleration,
headway, spacing, etc. The observations in this chapter are the empirical basis for
the book.

Chapter 4
A Markov Model Based on Headway/Spacing Distributions

4.1 Introduction

The stationary headway/spacing distribution models in Chap. 2 have been often criticized for neglecting the dynamic role of traffic (Bovy 2001). A modern view accepted by many researchers is: the explicit distribution observed in practice should be a reflection of the implicit interaction between vehicles. The fact that some stationary distribution models are only suitable for free flow is because the interactions between consecutive vehicles are relatively weak and neglectable in free flow.

In recently developed dynamic headway distribution models (Abul-Magd 2007; Krbalek 2007; Krbalek and Helbing 2004; Krbalek and Seba 2009; Krbalek et al. 2001; Mahnke and Kühne 2007; Nishinari et al. 2003; Treiber et al. 2006b) traffic is described as N strongly-linked particles under fluctuations. Notice that the governing interaction forces or potentials are not directly measurable for traffic applications; the statistical distributions of particles are investigated instead. For example, in Krbalek (2007), Krbalek and Helbing (2004), Krbalek and Šeba (2009), Krbalek et al. (2001), Random Matrix Theory was applied to predict headway distributions among vehicles in different phases of traffic flows. These studies revealed why the shapes of headway/spacing distributions did not change too much, even in different traffic states (free flow, congested flow).

However, previous statistic approaches mainly addressed on the steady-state of the traffic; how to depict the transient-state of inter-arrival/inter departure vehicle queuing interactions still needs further discussions.

A natural idea to answer the above questions is to develop microscopic simulation models, e.g., car-following, Cellular Automata (CA), which meanwhile yield such statistical properties. In Krbalek and Helbing (2004), Mahnke and Kühne (2007), Nishinari et al. (2003), Treiber et al. (2006b), different microscopic car-following models were proposed. For example, in Treiber et al. (2006b), a special stochastic repulsive interaction-based car-following model was reformulated into a Fokker-Planck type equation, which calculated the steady-state space-gap distribution among

© Tsinghua University Press, Beijing and Springer-Verlag Berlin Heidelberg 2015
X. (M.) Chen et al., *Stochastic Evolutions of Dynamic Traffic Flow*,
DOI 10.1007/978-3-662-44572-3_4

the elements in a queue as a function of their interaction potentials. However, the obtained formulas of the unseen potential that governs the movements of the vehicles are not directly interpreted by our ordinary driving experiences.

In this chapter, we link two research directions of road traffic: mesoscopic headway distribution model and microscopic vehicle interaction model, together to account for the empirical headway/spacing distributions. A unified car-following model will be proposed to simulate different driving scenarios, including traffic on highways and at intersections. The parameters of this model are directly estimated from NGSIM trajectory data. In this model, the empirical headway/spacing distributions are viewed as the outcomes of stochastic car-following behaviors and also the reflections of the unconscious and inaccurate perceptions of space and/or time intervals that people may have. This explanation can be viewed as a natural extension of the well-known psychological car-following model (action point model). Besides, the fast simulation speed of this model will benefit transportation planning and surrogate testing of traffic signals.

To explain the phenomenon that observed headways follow a certain lognormal type distribution within each preselected velocity range, an asymmetric stochastic extension of the well-known Tau Theory is proposed. It assumes that the observed headway distributions come from drivers' consistent actions of headway adjusting, and more importantly, the intensity of headway change is proportional to the magnitude of headway on average. The agreement between the model predictions and the empirical observations indicates that the physiological Tau characteristics of human drivers govern driving behaviors in an implicit way.

4.2 A Markov Model for Headway/Spacing Distributions

4.2.1 Background

In this section, we will summarize and further improve the results on the following issues:

(1) The first problem left unsolved in the models (Jin et al. 2009; Li et al. 2010a; Wang et al. 2009a) is the parameters of the certain stochastic processes are indirectly estimated from the steady-state headway/spacing distributions. To achieve a solid proof, we use the NGSIM trajectory data to directly set the parameters. Testing results will show that the above conjecture on stochastic processes-based car-following modelling is correct.

(2) The second problem discussed in Jin et al. (2009), Li et al. (2010a), Wang et al. (2009a) is whether we can set up a unified simulation model for different driving scenarios, including highway and road intersections. From the viewpoint of the vehicles' queuing dynamics, we need to depict four kinds of traffics: (a) free-flows formulating no explicit queues; (b) stable moving queues (vehicle platoons); (c) unstable queues, which contains complex inter-arrival and inter-departure queuing interactions; (d) static queues (e.g., a line of vehicles parked roadside, vehicle queues fully stopped before a signalized intersection).

Tests in Jin et al. (2009), Li et al. (2010a), Wang et al. (2009a) showed that we could view the varying process of the headway/spacing between two consecutive vehicles as a certain stochastic process, whose steady-state distribution corresponded with the empirical distribution directly. Moreover, such stochastic phenomena can be naturally explained as the outcomes of stochastic driving behaviors, since a human driver can neither constantly maintain a desired velocity nor accelerate or decelerate in a smooth fashion (Chowdhury et al. 2000; Helbing 2001; Mahnke et al. 2005). However, most car-following models (Aycin and Benekohal 1999; Brackstone and McDonald 1999; Mehmood et al. 2003; Panwai and Dia 2005) emphasize on other performance indices for traffic simulation. Moreover, it explicitly incorporates randomness in the dynamic model (just like many CA models of road traffic in Maerivoet and De Moor (2005)), which makes it easy to reproduce other statistics of traffic flow. But unlike CA models, the space continuity of this model affords more accuracy in fitting the empirical distributions than those cellular automata, which also considers probabilistic headways/spacings (Hu et al. 2008; Li et al. 2008).

In the discussions of Jin et al. (2009), Li et al. (2010a), Wang et al. (2009a), we found such models could be extended to characterize the statistics of unstable queues). In Jin et al. (2009), the departure headways of a dissipating vehicle queue at the intersection (when the traffic light turns green) were carefully studied. It was found that the departure headway distribution at each position of a dissipating queue had a similar shape (lognormal distribution or more generic distribution alike) to the headways observed on highway. This triggered us to use a unified simulation model to reproduce the departure headways at intersections as well as the headways on highway. Simulation results proved the possibility of such approaches. The entering spacings at signalized intersections were studied, where entering spacing was defined as the spacial distance between two successive vehicles fully stopped in a traffic lane, when the traffic light turned red.

The problem of parking spacings had received considerable attentions since Renyi's model Renyi (1963). Different models based on *random sequential adsorption* and *random matrix theory* had been proposed during the last four decades (Lee 2004; Rawal and Rodgers 2005; Talbot et al. 2000). One interesting and intuitive model had recently been proposed in Petr (2008), in which the empirical spacing distributions could also be generated as the outcome of a Markov type adjusting process of the spacings. This could be viewed as an extension of the models proposed in Wang et al. (2009a).

(3) The third problem is which variable (headway, spacing or velocity) should be chosen as the fundamental random variable. Some approaches emphasized the distributions of velocities, but more microscopic simulation models assumed that drivers adopt relatively constant time headways. Based on the NGSIM trajectory data, we show that this assumption is not accurate. Indeed, we find that (a) when the velocity is low, the drivers will try to maintain a certain safe spacing whose distribution is relatively constant; (b) when the velocity is high, the drivers prefer to maintain a certain headway, whose distribution belongs to a lognormal like distribution family. And the psychological car-following model can be extended to account for this interesting phenomenon.

4.2.2 Markov-Process Simulation Models

The empirical results indicate that we should choose headway as the fundamental variable in simulation when velocity is high, while spacing when velocity is low. The interesting finding also raises the following question:

Question 4.1. Why can we observe similar statistics during different traffic states? Is there any microscopic explanation behind?

In some recent approaches (Jin et al. 2009; Li et al. 2010a; Wang et al. 2009a), the observed distributions were viewed as the results of consistent headway/spacings adjusting actions. Because of the unconscious and also inaccurate perceptions of space and/or time interval that people have, these adjusting actions are stochastic and discrete in time. This new viewpoint brings at least two benefits: (a) we can then regard the consistent adjusting actions as a certain stochastic process, whose steady-state distribution may be directly calculated and compared with the empirical ones; (b) this explanation is in accordance with the psychological car-following model (action point model), which had been well discussed in the 1970–1990s (Evans and Rothery 1977; Hancock 1999; Michaels 1963; Ranney 1999).

In this section, we will first review some previous models and then propose the Markov-headway model for high velocity and the Markov-spacing model for low velocity. The associated psychological explanation is also presented.

4.2.2.1 Markov-Spacing Model

In Wang et al. (2009a), the varying process of the spacing during following was assumed to be a Markov process. Since it is difficult to deal with continuous Markov process in simulation, we assume altogether n possible aggregated spacing states, each of which represents a certain range of spacing.[1] To discretize the continuous spacing into several nonoverlapping states is a simplification to introduce the Markov transfer probability. We can define a certain transition kernel by applying continuous-state Markov process, but it will make the simulation too difficult to implement.

Considering the relatively long time of driving actions of many drivers, the transition probability from state i to state j at a certain time would approach a steady value $P_{ij}, i, j = 1, 2, ..., n$. Based on the property of Markov process, given a transition matrix P, we can directly get the aggregated steady-state spacing distribution. This model was shown to successfully explain and reproduce the entering spacings (Wang et al. 2009a). However, the transition matrix P was not directly estimated from the car-following data. Indeed, a guess of P was constructed in Wang et al. (2009a) to allow a fastest convergence speed of the obtained Markov process.

We can see that when the velocity is low, drivers prefer to maintain a certain safe space gap whose distribution is relatively constant. In such scenarios, we should apply the Markov-spacing model. Because there are not enough low-velocity spacing

---

[1] The aggregation technique is a useful tool in Markov processes modeling and analysis. It removes some unnecessary details from the original complete Markov process and generates a simpler model still with good approximation accuracy.

data in NGSIM, we do not study the calibration of the transition matrix for Markov spacing model in this chapter.

In Wang et al. (2009a), simulations show that the entering spacings could be viewed as an approximation of the saturation spacing and can be reproduced by adopting the above Markov-spacing model. It is further shown in Jin et al. (2009) that the distribution of entering spacings was different from that of the saturation. This is because: (a) vehicles are allowed to move back and forth when parking roadside but not before intersections; (b) more importantly, the governing stochastic dynamic for these two spacing varying processes are different.

4.2.2.2 Markov-Headway Model

Inspired by the previous approaches, we propose a Markov-headway model for high-way traffic simulation.

Proposition 4.1 *In a finite dimension of state space* $\mathbb{N} = \{1, 2, ..., n\}$, *the transition probability from states i to j for headway is $P_{ij} \in (0, 1)$, satisfying $\sum_j P_{ij} = 1$, $\forall i, j \in \mathbb{N}$. The corresponding transition probability matrix $\boldsymbol{P}$ can be estimated by the observation samples $h_k(t) \rightarrow h_k(t + \Delta t)$ according to central limit theorem when the sample number is large enough*

$$\boldsymbol{P} \approx (P_{ij}), \quad P_{ij} = \mathcal{J}_{ij}/\mathcal{J} \tag{4.1}$$

where $\mathcal{J}_{ij}$ is the number of samples that $h_k(t)$ belongs to state i while $h_k(t + \Delta t)$ belongs to state j, $\mathcal{J}$ is the population.

Noticing that headway distributions vary with velocity, the process is depicted by three Markov processes according to different velocity ranges[2]: 0–5, 5–10, and 10–15 m/s. Because there are few records with velocity greater than 15 m/s in the NGSIM trajectory data of the US Highway 101 site, we simply discard these rare records.

Clearly, we can divide even more velocity ranges to achieve even better simulation results. Here, to use three Markov processes is a trade-off between simulation speed and accuracy. For each velocity range, we aggregate the sample headways h_i into 10 possible states,[3] namely 0–1, 1–2, ..., 9–10 s. The headways that are larger than 10 s are neglected, since it indicates the vehicle is in a free driving state.

Table 4.1 shows the transition probability matrices $\boldsymbol{P} = (P_{ij})$ for each velocity range. We can thus calculate the steady-state distribution of each Markov processes from

$$\boldsymbol{\pi}_\ell = \boldsymbol{\pi}_\ell \boldsymbol{P}_\ell, \quad \ell = 1, 2, 3 \tag{4.2}$$

---

[2] Actually, this division plan implies to the division of free flow, congestion flow, and the middle states.

[3] It is allowed to choose nonuniform aggregation length here. Since the aggregation method is not our main focus, we neglect the related discussions here.

Table 4.1 The estimated transition matrices, $v \in [0, 5) \cup [5, 10) \cup [10, 15]$ m/s

P_{ij}	1	2	3	4	5	6	7	8	9	10
1	0.5000	0.5000	0.0000	0.0000	0.0000	0.0000	0.0000	0.0000	0.0000	0.0000
2	0.0341	0.4205	0.4318	0.0682	0.0341	0.0114	0.0000	0.0000	0.0000	0.0000
3	0.0000	0.0505	0.6291	0.2456	0.0466	0.0155	0.0058	0.0039	0.0029	0.0000
4	0.0000	0.0067	0.2345	0.5529	0.1345	0.0395	0.0176	0.0101	0.0034	0.0008
5	0.0000	0.0015	0.0507	0.3209	0.4448	0.1179	0.0358	0.0134	0.0104	0.0045
6	0.0000	0.0000	0.0168	0.1148	0.3249	0.3557	0.1232	0.0336	0.0196	0.0112
7	0.0000	0.0000	0.0108	0.0919	0.1784	0.2000	0.3351	0.0973	0.0757	0.0108
8	0.0000	0.0000	0.0227	0.0455	0.1591	0.2159	0.0909	0.2614	0.1477	0.0568
9	0.0000	0.0000	0.0172	0.0345	0.1207	0.1724	0.1207	0.1034	0.2759	0.1552
10	0.0000	0.0000	0.0000	0.0667	0.0333	0.2333	0.0667	0.0667	0.1667	0.3667
P_{ij}	1	2	3	4	5	6	7	8	9	10
1	0.4787	0.5106	0.0106	0.0000	0.0000	0.0000	0.0000	0.0000	0.0000	0.0000
2	0.0047	0.8524	0.1404	0.0017	0.0004	0.0000	0.0000	0.0000	0.0001	0.0001
3	0.0000	0.1087	0.8248	0.0633	0.0023	0.0002	0.0003	0.0002	0.0002	0.0000
4	0.0000	0.0009	0.2179	0.7090	0.0688	0.0031	0.0003	0.0000	0.0000	0.0000
5	0.0000	0.0000	0.0061	0.2670	0.6323	0.0850	0.0097	0.0000	0.0000	0.0000
6	0.0000	0.0000	0.0000	0.0286	0.3250	0.5571	0.0893	0.0000	0.0000	0.0000
7	0.0000	0.0000	0.0000	0.0000	0.0098	0.2941	0.6078	0.0784	0.0098	0.0000
8	0.0000	0.0000	0.0000	0.0000	0.0000	0.0204	0.2653	0.5714	0.1224	0.0204
9	0.0000	0.0000	0.0000	0.0000	0.0909	0.0000	0.0909	0.1818	0.6364	0.0000
10	0.0000	0.0000	0.0000	0.0000	0.0000	0.1250	0.0000	0.1250	0.2500	0.5000
P_{ij}	1	2	3	4	5	6	7	8	9	10
1	0.7547	0.2421	0.0011	0.0000	0.0011	0.0011	0.0000	0.0000	0.0000	0.0000
2	0.0149	0.9271	0.0574	0.0004	0.0001	0.0001	0.0000	0.0000	0.0000	0.0000
3	0.0000	0.0888	0.8679	0.0431	0.0001	0.0001	0.0000	0.0000	0.0000	0.0000
4	0.0000	0.0003	0.1303	0.8198	0.0489	0.0006	0.0000	0.0000	0.0000	0.0000
5	0.0000	0.0012	0.0012	0.1673	0.7651	0.0652	0.0000	0.0000	0.0000	0.0000
6	0.0000	0.0000	0.0000	0.0033	0.1759	0.7492	0.0684	0.0033	0.0000	0.0000
7	0.0000	0.0000	0.0000	0.0000	0.0095	0.1810	0.7333	0.0762	0.0000	0.0000
8	0.0000	0.0000	0.0000	0.0000	0.0000	0.0000	0.2400	0.6000	0.1200	0.0400
9	0.0000	0.0000	0.0000	0.0000	0.0000	0.0000	0.0769	0.2308	0.6154	0.0769
10	0.0000	0.0000	0.0000	0.0000	0.0000	0.0000	0.0000	0.0000	1.0000	0.0000

where $\boldsymbol{\pi}_\ell$ is the stationary distribution vector of headway for the ℓth velocity range, $\boldsymbol{P}_\ell$ is the corresponding transition probability matrix for the ℓth velocity range, ℓ corresponds to $0 \leqslant v < 5$, $5 \leqslant v < 10$ and $10 \leqslant v < 15$ m/s, respectively, whose sample sizes are 3,698, 20,943 and 32,475. If $\boldsymbol{\pi}_\ell$ exists, then $\boldsymbol{P}_\ell^\infty = [\boldsymbol{\pi}_\ell, \boldsymbol{\pi}_\ell, ..., \boldsymbol{\pi}_\ell]^T$.

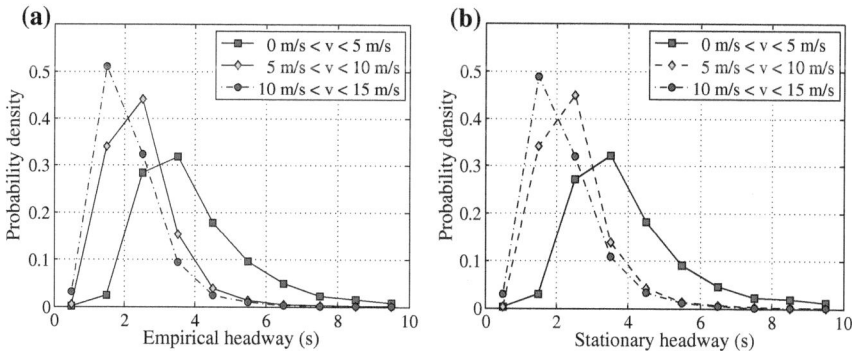

Fig. 4.1 The discrete distribution of **a** the aggregated empirical headways based on the samples and **b** the steady-state headways obtained from the observed transition matrices

As shown in Fig. 4.1, these calculated distributions agree well with the empirical distribution. This fact verifies our conjecture that we can model car-following actions by using Markov processes.[4]

Suppose the locations of the $(n-1)$th vehicle and nth vehicle at time t are denoted as $x_{n-1}(t)$ and $x_n(t)$, respectively. And the associate velocities are $v_{n-1}(t)$ and $v_n(t)$. The spacing is denoted as $s_{n,n-1}(t) = x_{n-1}(t) - x_n(t)$. The headway is written as $h_{n,n-1}(t) \approx s_{n,n-1}(t)/v_n(t)$.

In order to better depict the constraints of vehicle dynamics and human maneuvers, we divide the driving scenarios into four phases as shown in Fig. 4.2. At each simulation time t, the nth vehicle will determine which mode should be applied:

Fig. 4.2 Car-following phase-space diagram for the proposed model

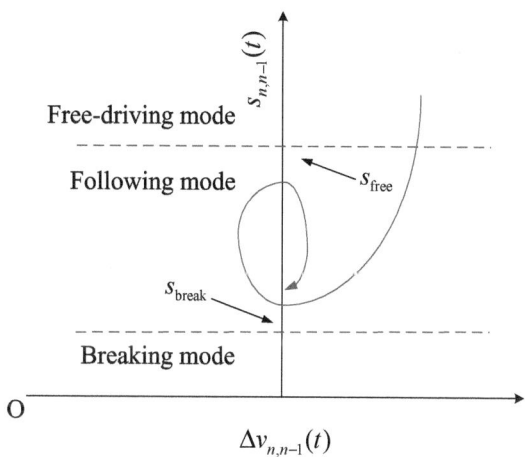

---

[4] A benefit of Markov model is that we can fit almost any kind of empirical distributions. In other words, we can allow the steady-state distribution to be the more generic Γ distribution.

1. *Free driving mode*: When the spacing is larger than a preselected maximum threshold s_{free}, the vehicle is in free driving mode. It will try to reach and maintain the highest possible velocity v_{max} as

$$v_n(t + \Delta t) = \min \{v_{\text{max}}, \ v_n(t) + a_{\text{max}} \Delta t\} \tag{4.3}$$

where a_{max} is the maximum acceleration rate.

2. *Starting mode*: When the nth vehicle fully stops at time t and the spacing $s_{n,n-1}(t)$ is larger than a preselected minimum threshold s_{start}, the vehicle will try to speed up as

$$\begin{cases} \Pr\{v_n(t + \Delta t) = a_{\text{max}} \Delta t\} = p_{\text{slow}} \\ \Pr\{v_n(t + \Delta t) = 0\} = 1 - p_{\text{slow}} \end{cases} \tag{4.4}$$

This mode is introduced to mimic the slow-to-start property of human drivers.[5]

3. *Breaking mode*: If the spacing $s_{n,n-1}(t)$ is smaller than a preselected minimum threshold s_{break}, the vehicle will stop as

$$v_n(t + \Delta t) = 0 \tag{4.5}$$

4. *Following mode*: When the spacing satisfies $s_{\text{stop}} < s_{n,n-1}(t) < s_{\text{free}}$ at time t, the nth vehicle is in following mode. We will apply the kernel Markov-headway car-following model as follows:

Step 1. Suppose at time t, headway $h_{n,n-1}(t)$ falls into state i as $h_{n,n-1}(t) \in (H_i^-, H_i^+]$. Thus, at time $t + \Delta t$, the desired headway $\tilde{h}_{n,n-1}(t + \Delta t)$ is determined as

$$\Pr\{\tilde{h}_{n,n-1}(t + \Delta t) = h_{\text{random}}\} = P_{ij}, \ \forall i, j \in \mathbb{N} \tag{4.6}$$

where h_{random} is a randomly generated number, which has a uniform distribution $h_{\text{random}} \sim \mathcal{U}(H_j^-, H_j^+]$.

The desired nth vehicle's velocity is approximately formed as

$$\tilde{v}_n(t + \Delta t) \approx s_{n,n-1}(t) / \tilde{h}_{n,n-1}(t + \Delta t) \tag{4.7}$$

Step 2. A safe velocity lower bound $v_{n,\text{safe}}(t + \Delta t)$ at time t is also calculated to guarantee collision free at time $t + \Delta t$ as

$$s_{n,n-1}(t) + \left(v_{n-1}(t) - \frac{v_n(t) + v_{n,\text{safe}}(t + \Delta t)}{2} \right) \Delta t \geqslant s_{\text{brake}} \tag{4.8}$$

In other words, we have

---

[5] Another frequently mentioned driving feature, random deceleration, has been directly embedded in the proposed model as the Markov-type transitions from a smaller headway to a larger headway.

$$v_{n,\text{safe}}(t + \Delta t) = \max \left\{ 0, \ \frac{2(s_{n,n-1}(t) - s_{\text{brake}})}{\Delta} + v_{n-1}(t) + \Delta v_{n,n-1}(t) \right\} \quad (4.9)$$

where $\Delta v_{n,n-1}(t) = v_{n-1}(t) - v_n(t)$.

Step 3. If $\tilde{v}_n(t + \Delta t) > v_n(t)$, the final velocity at time $t + \Delta t$ is chosen as

$$v_n(t + \Delta t) = \min \left\{ \tilde{v}_n(t + \Delta t), \ v_n(t) + a_{\max} \Delta t, \ v_{\max}, \ v_{n,\text{safe}}(t + \Delta t) \right\} \quad (4.10)$$

If $\tilde{v}_n(t + \Delta t) \leqslant v_n(t)$, the final velocity at time $t + \Delta t$ is chosen as

$$v_n(t + \Delta t) = \min \left\{ \tilde{v}_n(t + \Delta t), \ \max \left\{ v_n(t) - b_{\max} \Delta t, \ 0 \right\}, \ v_{n,\text{safe}}(t + \Delta t) \right\} \quad (4.11)$$

where $b_{\max}$ is the maximum deceleration rate.

When the velocity is determined, the location, spacing and headway of the nth vehicle are updated as

$$\begin{cases} x_n(t + \Delta t) \approx x_n(t) + [v_n(t) + v_n(t + \Delta t)]\Delta t/2 \\ s_{n,n-1}(t + \Delta t) = x_{n-1}(t + \Delta t) - x_n(t + \Delta t) \\ h_{n,n-1}(t + \Delta t) \approx s_{n,n-1}(t)/v_n(t + \Delta t) \end{cases} \quad (4.12)$$

4.2.2.3 Psychological Explanations

As assumed in van Der Hulst (1999), Fuller (2005), Taylor (1964), van Winsum (1999), Wilde (1982), drivers usually used time headway as a safety margin. When drivers were required to follow at a headway smaller than preferred, his anxiety would significantly increase, which could be measured as a reduction in heart rate variability.

In Taylor (1964), it was argued that drivers attempted to maintain a constant level of anxiety when driving, which sometimes was interpreted as that drivers will keep a constant headway. But based on the statistics, it is better to say that drivers will maintain a constant headway when the velocity is high and a constant spacing when the velocity is low.

In other words, this human decision process can be taken as the subjective estimation of the probability of collision (Wilde 1982). We can also adopt the explanation in van Der Hulst (1999), in which drivers would maintain a level of task difficulty. When the velocity is high, to keep a relatively constant headway is the dominant task; when the velocity is low, to reserve a certain spacing becomes more important.

As pointed out in many literatures (Fuller 2005; van Der Hulst 1999; van Winsum 1999), human drivers could not maintain a strictly constant anxiety (headway/spacing) due to several reasons (including physiological unsteadiness, perception errors, and mechanical constraints of vehicles). But few action models discuss how to model the fluctuations of headway or spacings.

In this section, we argue that such deviations of the desired headways/spacings can be modeled by Markov processes. The abstraction of the continuous headway into several discrete states could be viewed as a conceptual usage of noticeable change in the psychological car-following models, where the boundaries of the states in Markov models can be taken as a kind of thresholds of visual perception.

The leading vehicle's movement is often unpredictable (at least not fully predictable), the accelerating and braking action of the driver is therefore often overdue. As a result, drivers tend to take very careful acceleration when the spacing between the two vehicles is small and do not want to speed up at once after braking. From the viewpoint of Markov transition, we observe higher probability to transfer to a state with larger headway or spacing, when the current state indicates a headway shorter than mean headway (preferred headway); and on the contrary lower probability to transfer to a state with smaller headway/spacing, when current headway is larger than preferred. The coincidences of the empirical distributions and the steady-state distributions got from the observed transition matrices are strong proofs for the Markov model argument.

4.2.3 Simulation Results

The above discussions answer Question 4.1. We will present some simulation results to show: (a) the proposed model can generate the headway/spacing distributions as observed; (b) it can fulfill other performance indices (e.g., low spacing and velocity tracking errors) that had been used to evaluate car-following models (Aycin and Benekohal 1999; Brackstone and McDonald 2007; Cassidy and Windover 1998; Ossen and Hoogendoorn 2005; Panwai and Dia 2005).

4.2.3.1 Distribution Test of the Markov-Headway Model

To simulate headway distributions for highway traffic, a 10,000 m cycling single-lane road system with 500 vehicles is constructed in the simulation test. The parameters are $v_{max} = 15$ m/s, $a_{max} = 3$ m/s^2, $b_{max} = 4.5$ m/s^2, $l = 3.5$ m, $s_{brake} = 5$ m, $s_{free} = 50$ m, $s_{start} = 10$ m, $p_{slow} = 0.4$.

To achieve better simulation results, we divide the headways into 20 possible states, i.e., 0–0.5, 0.5–1, ..., 9.5–10 s, with respect to three velocity ranges of 0–5, 5–10, and 10–15 m/s. The corresponding transition matrices are still obtained from the NGSIM trajectory data directly.

Initially, the vehicles are supposed to be uniformly 20-meter apart from each other and the initial velocity is $v_0 = 10$ m/s. The velocity, headway, spacing of each vehicle are recorded during the 1,200 s simulation. As pointed out in Sugiyama et al. (2008), we can observe free flow, congestion flow, and the middle states in such a cycle road without bottlenecks. It is convenient to collect simulated headways in different traffic states.

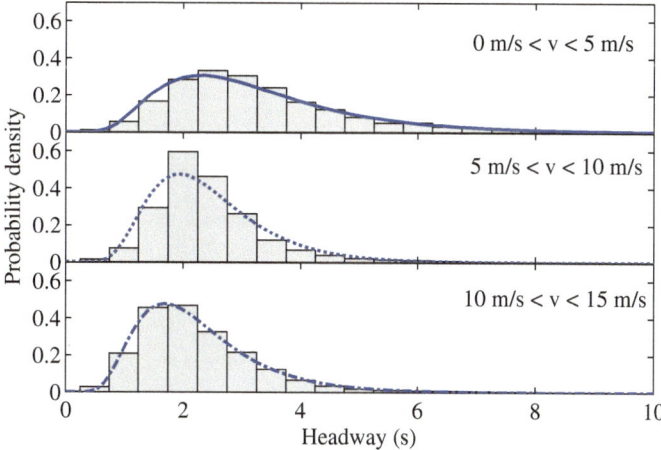

Fig. 4.3 The simulated headways and the fitted lognormal distributions

Figure 4.3 shows the corresponding distributions of the simulated headways within different velocity ranges. This proves that the proposed model yields similar headway distributions as the observations.

4.2.3.2 Tracking Performance Test of the Markov-Headway Model

We also explore how this new model replicates the relative velocity and spacing between the leader and follower vehicles. In this test, only two successive vehicles are considered. The velocity and acceleration profile of the leader is ported from NGSIM dataset. Only the dynamics of the follower is simulated and compared with the values from NGSIM.

To avoid the influences of lane changing behaviors, only the trajectory data of the vehicles forming a fixed platoon (containing more than three consecutive vehicles all the way during observation) from NGSIM are retrieved and used. Figure 4.4 shows the trajectories of five groups of such vehicles.

Figure 4.5 shows some typical results of the relative velocity and spacing differences. It is clear that the outputs of the proposed model agree with the real ones. As suggested in Panwai and Dia (2005), we check the following performance criteria:

(a) The Root Mean Square Error (RMSE) metric on spacing

$$\text{RMSE} = \sqrt{\frac{1}{MT} \sum_{m=1}^{M} \sum_{t=1}^{T} (d_s(t) - d_e(t))^2} \qquad (4.13)$$

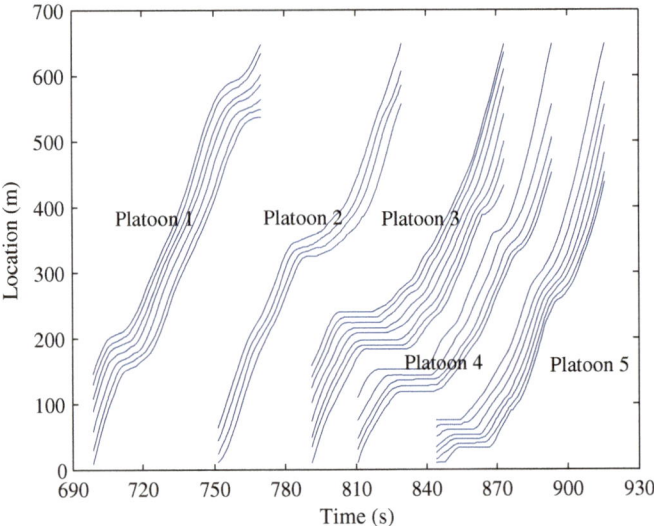

Fig. 4.4   Trajectories of the selected 5 platoons of consecutive vehicles

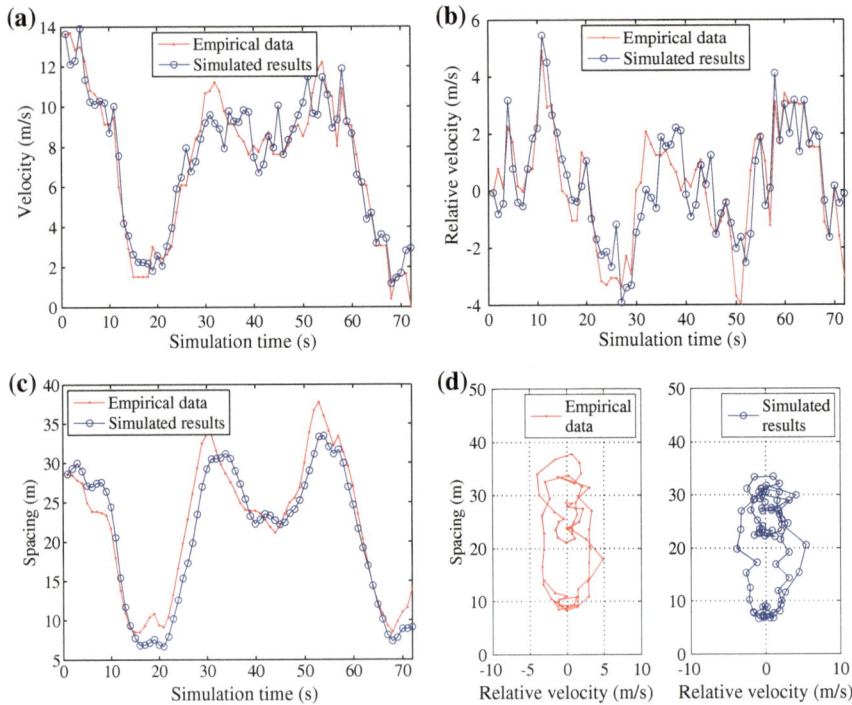

Fig. 4.5   Comparison of field data with the proposed stochastic car-following simulation

Table 4.2 Performance of the car-following model in the traching tests

Platoon	1	2	3	4	5
RMSE	10.77	8.50	8.68	7.13	9.21
EM	0.51	0.39	0.45	0.40	0.37
Observations	432	237	581	332	432
Vehicles	7	4	8	5	7

where $d_s(t)$ is the simulated neat spacing at a certain time t, $d_e(t)$ is the field neat spacing at the same time, T represents the total observation times, M represents the number of examined trajectories.

(b) The Error Metric (EM) on spacing

$$\text{EM} = \sqrt{\frac{1}{MT} \sum_{m=1}^{M} \sum_{t=1}^{T} \left(\log \frac{d_s(t)}{d_e(t)}\right)^2} \qquad (4.14)$$

where the error is weighted by the logarithm and squared to avoid overrating discrepancies for large distance.

In the simulation, $T = 75\,$s, and we can find totally $M = 25$ vehicles satisfied our fixed platoon requirements. Table 4.2 shows that the obtained RMSE and EM values are generally equivalent to many other important car-following models (Panwai and Dia 2005). This proves the effectiveness of this new model.[6]

Simulations also reveal that the aforementioned model yields relatively larger spacing errors when velocity is low (e.g., see the curves of empirical and simulated spacings from $t = 30$ to $t = 60\,$s in Fig. 4.5c). This is mainly because the braking rule is an oversimplification here.

Thus, to achieve better simulation results, we shall develop a unified model, which appropriately combines the Markov-headway mechanism (when driver is starting up, approaching, or unconsciously following) and the Markov-spacing mechanism (when driver is stopping or braking) simultaneously. However, there are few low-velocity spacing data in NGSIM dataset; thus, we will not further discuss how to build such a model in this chapter.

4.2.4 Discussions

- **Model calibration**

  The parameters in the above Markov headway model are determined in two ways:
  (a) the transition matrix P is directly estimated from the observed transition frequencies; (b) the other parameters (e.g., $v_{\max}$) are carefully simplified from other

---

[6] It should be pointed out that different trajectory data are used in this section and Panwai and Dia (2005), which may influence the values of RMSE and EM, too.

empirical data. Simulations show that P is the dominant factor governing the distributions of headways/spacings. We can approximately reproduce the desired headways/spacings when P is correctly set, even if other parameters are not so accurate. We will not discuss how to calibrate other parameters, since this problem have received adequate discussions in the literature (Kesting 2008; Ossen 2008; Ossen and Hoogendoorn 2005; Ossen et al. 2007).

- **Simulation time interval/span**
  The simulation time interval Δt is assumed to be 1 s. We believe it is an appropriate value to allow the possible actions of human drivers and meanwhile reduce the incremental errors when updating locations. We can also set other value of time interval, as long as Δt is not so large. As follows, we need to reestimate P, since we indeed calculate the observed transition frequencies in one second. We also need to reset the other time-dependent model parameters, such as $a_{\max}$ and $b_{\max}$.

- **Reaction time/delay**
  Reaction time is a characteristic of human drivers in vehicle manipulation. How to model and calibrate reaction time is an important problem in car-following models (Kesting 2008; Ma and Ingmar 2006; Ossen 2008). In the proposed model, the reaction delay is implicitly embedded in a stochastic manner: the following drivers has a nonzero probability to maintain the current velocity even when the velocity of the leading vehicle has been changed. We can also take it as a randomly distributed reaction delay in the proposed model. Moreover, we combine the uncertainness and the time delay into the randomness of headway variations. This is a trick that is frequently used in CA models of road traffic (Hu et al. 2008; Li et al. 2008; Maerivoet and De Moor 2005).

In short, we establish a new car-following model to fulfill the following considerations: (a) it yields the steady-state headway/spacing distributions as those are observed in practice. Moreover, it is a microscopic simulation model so that it can simulate transient-state statistics of road traffic, too; (b) it provides a unified model to simulate different driving scenarios, including highway traffic and urban street traffic; (c) it is in accordance with our daily driving experiences. The stochastics explicitly embedded in this model could be reasonably explained as the outcome of the unconscious and also inaccurate perceptions of space and/or time interval that people have. Thus, this model is an extension of the famous psychological car-following model (action point model); (d) it is a fast simulation model for surrogate testing.

Constrained by the length of this section, the following issues are not emphasized: (a) in real traffic, different kinds of drivers (e.g., aggressive vs. passive, young vs. old, skilled vs. green-hand, rigorous vs. fatigued) are running different kinds of vehicles (e.g., cars vs. trucks, buses) on the same road and thus the traffic flow is heterogeneous. Field testing data have proved that various headways and spacings can be observed among different traffic participators. To deal with this problem, we had better estimate and assign distinct transition matrices for each kind of traffic participators. This naturally improves the transferability of the proposed models; (b) the traffic control measures and visibility condition (Broughton et al. 2007) should also be considered when collecting headways/spacings in practice. As discussed in Michaels and Solomon (1962), even a speed sign can notably alter the value of mean headway. Therefore, the impact of such environmental factors should be carefully handled in elaborate traffic simulation and planning; (c) the proposed model can also reproduce many complex phenomena of highway traffic flows, including wide moving jam, pinned localized cluster, stop and go waves, and oscillated congested traffic; (d) the lane changing/merging behaviors are not discussed in this section. But the above models provide a basis to further incorporate the well-known stochastic gap acceptance model (Choudhury 2007; Daganzo 1981; Mahmassani and Sheffi 1981).

4.3 Asymmetric Stochastic Tau Theory in Car-Following

The statistics of time headway have gained a lot of attention in the field of transportation engineering for its significant impacts on traffic signal timing (Transportation Research Board of the National Academies 2010), traffic flow physics, and behavioral studies. It is often viewed as a reflection of the implicit forces that govern human's driving (see Brackstone and McDonald 2007; Brackstone et al. 2009; Ranney 1999; van Winsum 1999).

In recent years, a number of experiments have been conducted in an effort to examine headway adjustment behavior in car-following (Banks 2003; Chen et al. 2010b; Taieb-Maimon and Shinar 2001). The results indicate that drivers tend to maintain a generally consistent headway while within a single specific driving scenario. However, it is impossible for a driver to maintain a fixed headway, because of the unpredictability of leading vehicle movements and the inaccuracy of human perceptions, which lead to errors in decision-making and action. Drivers have to constantly adjust their behavior in an attempt to maintain their preferred headway. This process, which depends on the physiological response of the driver, yields a lognormal type headway distribution (or more generally, a Γ distribution) within each preselected velocity range (Chen et al. 2010a, b; Jin et al. 2009).

Since the early 1950s, many models have been proposed in order to explain headway distribution patterns (e.g., Abul-Magd 2007; Chen et al. 2010b; Jin et al. 2009; Krbalek et al. 2001; Li et al. 2010a; Nishinari et al. 2003; Transportation Research Board of the National Academies 2010; Wang et al. 2009a; Yan et al. 2011). In most of these stochastic models, the proposed visual forces that control vehicle movements are difficult to interpret in our daily driving experiences. Thus, a question is naturally raised as:

Question 4.2. Is there any driver-centric factors behind such lognormal distributions?

By further examining empirical headway distributions, we recently realized that this phenomenon can be physiologically explained by extending the Tau Theory on perception and action.

The Tau theory originates from the study of optical variable Tau ($\mathcal{T}$), which is defined as the projected angular size θ of an object divided by the relative change rate $d\theta(t)/dt$. In other words, it describes the change of the incoming object's image on the retina (Lee 1976, 2009), i.e., $\mathcal{T} = \theta(t)/\theta'(t)$. Tau ($\mathcal{T}$) is often regarded as an invariant. It is veridical and independent of other variables, such as the distance and velocity of the incoming object (Yan et al. 2011).

In this section, we study the ratio $\mathcal{T}$ between the change in headway in the next time interval and the current headway. Although different values of $\mathcal{T}$ can be observed in each iteration of headway adjustment process due to the randomness of human behavior, we found that the values can generally be represented as $\mathcal{T}^-$ and $\mathcal{T}^+$ for acceleration and deceleration cases, respectively. It is also widely accepted that the coexistence of $\mathcal{T}^-$ and $\mathcal{T}^+$ is caused by asymmetric driving behaviors (Edie 1961; Forbes 1963; Yeo 2008; Yeo and Skabardonis 2009).

The above findings indicate that microscopic driving behaviors influence macroscopic traffic flow dynamics in a latent and stochastic way. More importantly, we can identify intrinsic common physiological characteristics of driving behavior via the extrinsic measurements (e.g., headways).

4.3.1 Asymmetric Stochastic Extension of the Tau Theory

There are several measures of driving behavior characteristics. For example, Time-To-Collision (TTC) is a frequently mentioned one. It is defined as the time at which two vehicles would collide if both vehicles maintained their independent speeds and trajectories (Brackstone and McDonald 2007; Brackstone et al. 2009). That is, TTC is the ratio of the relative space gap divided by the relative speed Δv (a positive value represents a higher following vehicle speed).

Similarly but differently, Tau theory studies headway instead of TTC. The relationship between TTC and headway is $\text{TTC} = \Delta x/\Delta v = (hv - l)/\Delta v$ if $\Delta v > 0$, where Δx is the following gap, v is the following velocity, l is the vehicle length.

Similarly but differently, Tau theory studies time headway h instead of TTC. The original Tau Theory, which was developed by Lee (1976), stated that drivers usually focused on visual information, such as time-to-collision, instead of other types of information, such as space gap, speed, or ac/deceleration. More recently, Frost (2009) showed that all purposeful actions entailed controlling $\mathcal{T}$, which was the time-to-collision between the current position and the expected post-adjustment positions. Lee (2009) pointed out it was possible to build various $\mathcal{T}$ models depending on how the movements and states were defined.

From a neurophysiology perspective (see Farrow et al. 2006; Frost 2009; Schrater et al. 2001; Sun and Frost 1998), time headway change caused by driver actions is controlled by a parameter $\mathcal{T}$ as

$$\mathcal{T} \triangleq \frac{h(t)}{\Delta h(t)/T} = \frac{h(t)T}{h(t+T) - h(t)} \tag{4.15}$$

where $h(t)$ is the time headway at time t and $\Delta h(t) = h(t+T) - h(t)$ is the change of $h(t)$ within a defined time interval T, which is the time interval between two consecutive driver reactions.

If $\mathcal{T}$ is a constant, Eq. (4.15) indicates that the intensity of headway magnitude is proportional to headway change in T. Based on the famous Weber-Fechner Law,[7] Eq. (4.15) can also be explained as a linear function between visual stimulus (headway) and driver response (headway change) within a certain time interval T. This is in accordance with previous assumptions from psychophysical car-following models (Boer 1999; Brackstone and McDonald 1999; Michaels 1963). The difference is that previous psychophysical models aim to directly depict the visual angle subtended by the leading vehicle; while Eq. (4.15) attempts to incorporate perceptual reactions into headway change and provide an indirect measure for human visual control. However, we cannot directly fit Tau-theory model Eq. (4.15) with empirical observations. As discussed by Michaels (1963) and Boer Boer (1999), the perceptions and actions of human drivers are generally limited and inaccurate. In other words, $\mathcal{T}$ should not be a constant in the real world.

The randomness explicitly embedded in the evolution of headway can be explained as the unconscious and inaccurate perceptions of time interval and space that humans perform. If a driver is observed for a long enough time period, we can characterize the driver's car-following behaviors by examining the statistical features of this stochastic process. An advantage of this approach is that the steady-state distribution of this stochastic process can be directly calculated, and compared with empirical headway distributions as a way to validate the model. To model the randomness in driving behaviors, a special stochastic process was proposed in Jin et al. (2009), Li et al. (2010a) as follows:

$$\frac{h(t+T)}{h(t)} = \begin{cases} \beta & \Rightarrow p \\ 1/\beta & \Rightarrow 1-p \end{cases} \tag{4.16}$$

where $\beta \in (0, 1)$ is the scaling coefficient and $p \in (0, 0.5]$ denotes the tendency for the driver to reach a larger headway than preferred at time $t + T$. Specifically, only the situation is considered in Jin et al. (2009), Li et al. (2010a). Analogously, $1 - p$ is the tendency to reach a smaller headway than preferred at time $t + T$. As such, Eq. (4.16) can be viewed as a modified Tau Theory model, which can be rewritten as

$$\frac{\Delta h(t)}{h(t)} = \begin{cases} T/\mathcal{T}^- = \beta - 1 & \Rightarrow p \\ T/\mathcal{T}^+ = 1/\beta - 1 & \Rightarrow 1-p \end{cases} \tag{4.17}$$

or equivalently

---

[7] Weber, E. H. is one of the earliest researchers who quantitatively studied the relationship between the magnitude of stimuli and the perceived intensity of the stimuli. Then, Fechner, G. T. gave a math form of Weber's findings, which we now call the Weber-Fechner Law (Deco et al. 2007).

$$\mathcal{T} = \frac{h(t)}{\Delta h(t)/T} = \begin{cases} \mathcal{T}^- < 0 & \text{if } h(t+T) < h(t), \text{ with prob.} p \\ \mathcal{T}^+ > 0 & \text{if } h(t+T) > h(t), \text{ with prob.} 1-p \end{cases} \quad (4.18)$$

where $\mathcal{T}^-$ and $\mathcal{T}^+$ are the Tau factors, given that $(T/\mathcal{T}^- + 1)(T/\mathcal{T}^+ + 1) = 1$.

Equation (4.18) extends the original Tau Theory model of Eq. (4.15) in two ways:

- *Asymmetric behaviors*: a driver's behavior varies according to whether he/she is accelerating or decelerating. During acceleration, a driver will be more cautious and make smaller changes in headway. On the contrary, during deceleration, he/she will be less sensitive to the headway change. This asymmetry in vehicle acceleration and deceleration has been observed by many researchers (see Edie 1961; Forbes 1963). There are different ways to consider such asymmetry in car-following models. For example, Forbes Forbes (1963) explained this asymmetry as a difference in driver response time, which is slower during acceleration than deceleration. However, we consider more than just driver response time when addressing the asymmetric property of $\mathcal{T}$. Equation (4.18) does not indicate whether the Tau factors $\mathcal{T}^-$ and $\mathcal{T}^+$ are constant or velocity-dependent. This will be determined using field data in a later section.
- *Stochastic behaviors*: as reported in many literatures (e.g., Banks 2003; Brackstone and McDonald 2007; Chen et al. 2010b; Taieb-Maimon and Shinar 2001), drivers try to maintain a constant headway; but in reality, drivers maintain a stochastic headway that fluctuates around an expected value. In order to model this uncertainty, the probability p is introduced to characterize the tendency of reaching headway $h(t+T)$, which is smaller than the current headway $h(t)$. The first part of Eq. (4.16) asserts that, on average, drivers have a tendency/probability p to reach a smaller headway than preferred. It should be pointed out that Eq. (4.16) does not assert that drivers will adjust headway exactly as $\Delta h(t) = h(t)T/\mathcal{T}^-$ or $\Delta h(t) = h(t)T/\mathcal{T}^+$ for each t. However, if the statistical mean of $\Delta h(t)$ is calculated over a relatively long period, the adjustment behavior will lead to a headway change of $\Delta \bar{h}(t) = h(t)T/\mathcal{T}^-$. Similarly, we can interpret the second part of Eq. (4.16) as the tendency to reach a larger headway than preferred. In this case, the headway change can be expressed as $\Delta h(t) = h(t)T/\mathcal{T}^+$.

As proven in the literature (Jin et al. 2009; Li et al. 2010a; Limpert et al. 2001), a stochastic process, such as Eq. (4.16), would generate velocity-dependent lognormal headway distributions in simulations. However, differing from the previous physical analysis (Jin et al. 2009; Li et al. 2010a), the extended Tau Theory is a more penetrative and intrinsic explanation, because it is built on the assumption: *the limits of drivers' perceptions and actions will allow a certain $\mathcal{T}$-type proportional law between the headway and its increment* (Lee 2009). Equation (4.16) has not been previously validated by empirical flow data (Li et al. 2010a). To verify Eq. (4.16), we need to prove two assertions for $h(t) \rightarrow h(t+T)$:

- Drivers have a relatively constant tendency p to reach a smaller headway than preferred at different velocities; If this is ture, then

- Drivers have relatively constant $\mathcal{T}^-$ and $\mathcal{T}^+$ at different velocities and they satisfy that $\mathcal{T}^* = \left(T/\mathcal{T}^- + 1\right)\left(T/\mathcal{T}^+ + 1\right) = 1$.

Or equivalently,

Proposition 4.2 *In the sense of statistics, drivers tend to decrease expected headway with a relatively steady probability p under different velocities, in order to keep stationary $\mathcal{T}^-$ and $\mathcal{T}^+$, satisfying $\mathcal{T}^* = \left(T/\mathcal{T}^- + 1\right)\left(T/\mathcal{T}^+ + 1\right) = 1$.*

Notice that $h(t)$ is a continuous value. In order to accurately verify Eq. (4.15), it is necessary to estimate a transition kernel of the proposed stochastic process (Kallenberg 1997). However, this can be quite complicated and time-consuming. In addition, we may not have enough data to obtain acceptable results.

In this section, we use the aggregation method to estimate the statistics in Proposition 4.2. More precisely, these observed triples $(v_i(t), h_i(t), h_i(t + T))$ will be first divided into several distinct groups according to $v_i(t)$. Within each velocity group, the observed triples will be further divided into distinct subsets based on $\Delta h_i(t) = h_i(t + T) - h_i(t)$, where i denotes the sampling index. Finally, we examine the transition frequency among these subsets for each velocity group.

The headway adjustment time may be variable, but we cannot determine the exact adjustment time for each driver. So, we choose several discrete values of sampling time interval T to check its effects on the distributions of Tau values. Besides, we also need to further consider two dominant factors that may influence the estimation and validation of the extended Tau theory model: headway division strategies (N, i.e., headway domain is uniformly divided into N groups in order to specify headway change) and velocity change threshold (ϵ).

To find appropriate values for these factors, we choose the best parameter set from the following alternatives, i.e., $T = 0.5, 1.0, 2.0, 5.0$ s, $N = 5, 6, ..., 15$, and $\epsilon = 0.25, 0.5, 1.0, 2.0$ m/s. Though we can neither enumerate all the possible values of these factors nor analytically derive their optimal solutions, we believe that the empirical analysis would reveal the variation trend of the statistical results of p and $\mathcal{T}^*$.

To select the best values from the empirical alternatives, we propose two criteria:

1. *Mean criterion*: we compare the F statistic and p-value based on the analysis of variance (ANOVA) for each value of the aforementioned influencing factors. The smaller F statistic and the corresponding larger p-value (e.g., close to 1.00) indicate the hypothesis that the means of several populations are equal.
2. *Variance criterion*: we further compare the statistical features of p and $\mathcal{T}^*$ by using box-and-whisker diagram through five statistics, i.e., the lower quartile (25th percentile, Q_1), median (Q_2), upper quartile (75th percentile, Q_3), the smallest and largest observations within the interquartile ranges (i.e., $1.5 \times |Q_1 - Q_3|$). A smaller statistical range p and $\mathcal{T}^*$ for each value of influencing factors indicates a better concentration, and thus a better parameter set.

4.3.2 Testing Results

4.3.2.1 Data Preparation

To test Proposition 4.2 by using the NGSIM dataset (NGSIM 2006). A total number of 6,101 vehicles passed this road segment. In order to obtain nearly independent measurements, we set 12 virtual detectors every 50 m as stationary detectors, too, because extracting samples from the full trajectory information will lead to a high correlation among the data. Since the NGSIM dataset only monitors a 640 m freeway segment, we could not collect enough headway adjustment data on individual drivers from the available trajectory dataset. In this section, we examined the headway adjustment processes of the aggregate driver population. As such, the following statistics reflect the common characteristics of the aggregate driver population. This is also in accordance with the headway distribution measurements collected for the driver population, instead of individuals.

In this section, we conduct a three-step data filtering procedures for the NGSIM Highway 101 dataset:

Step 1. All data associated with lane changing maneuvers were discarded. The velocity and headway for each vehicle were recorded as the vehicle passed each virtual detector and at four later discrete points in time (i.e., 0.5, 1.0, 2.0, 5.0 s).

Step 2. We filter out the samples whose headways are larger than 10 s and spacings are larger than 50 m to retain car-following states. Data that showed velocities lower than 3 m/s were also filtered out to avoid error caused by vehicles that were fully stopped or travelling in a stop-and-go flow. When spacings are larger than 50 m, the vehicle will be running in free flow and drivers will pursue maximum velocity instead of safe headways.

Step 3. We use velocity of the following vehicle and velocity change of the leading vehicle to further filter the data. Those samples in which the leading vehicle velocity changes are larger than the threshold ε will also be discarded.

Table 4.3 shows the filtered samples of NGSIM Highway 101 dataset with the parameters $N = 10$ and $\epsilon = 1.0$ m/s as an example. After the three-step procedures of data filtering, we finally believe that the remaining samples reflect representative normal car-following behaviors.

Headway measurements are organized into 12 groups based on the recorded velocity, which ranges from 3 to 15 m/s. Each 1 m/s increase in velocity is considered as an individual velocity range group (i.e., 3–3.99 m/s; 4–4.99 m/s). Later, the headway is further divided into separate categories using a series of discrete values from 5 to 15. Let

$$\mathbb{S}_k \triangleq \left\{ \left(v_j(t), h_j(t), h_j(t+T) \right), \quad j = 1, ..., \mathcal{J}_k \right\}, \quad k = 1, 2, ..., 12 \quad (4.19)$$

where $\mathbb{S}_k$ is the kth set of triples, $\mathcal{J}_k$ is the number of records in the kth group.

Figure 4.6a shows the smoothed empirical joint probability density function of 56,481 filtered velocity-headway pairs from the NGSIM dataset. The obtained spacing data are unbiased because the measurement errors of space traveled is negligible.

Table 4.3 Headway adjustment data extracted from NGSIM Highway 101 dataset

T (s)	Data filtering	Sample size	Headway decrease	Headway increase	Percentage of head-way adjustment (%)
0.5	Step 1 and 2	16206	2252	2148	27
	Step 3	13827	1890	1789	27
1.0	Step 1 and 2	56481	10444	10154	36
	Step 3	42199	7614	7207	35
2.0	Step 1 and 2	15998	3647	4031	48
	Step 3	8465	1941	1930	46
5.0	Step 1 and 2	15675	4737	5206	63
	Step 3	4566	1507	1270	61

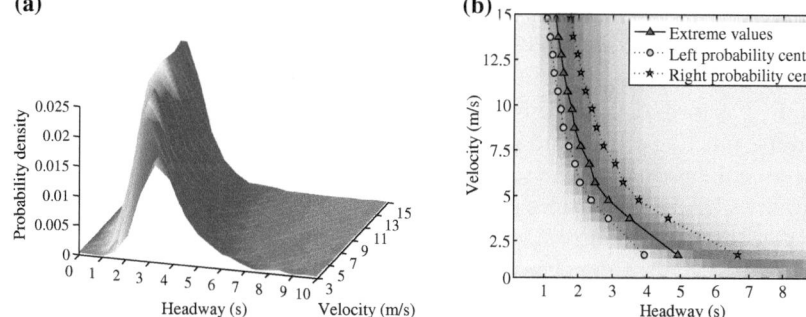

Fig. 4.6 Statistical results of empirical headway and velocity data. **a** Joint probability distributions of headway-velocity (h_i, v_i); **b** extreme values and centroids of marginal PDFs

A more detailed discussion on the analysis of measurement errors, acceleration, headway, and spacing during data processing and vehicle trajectory reconstruction can be found in Punzo et al. (2011). The measurement errors of velocities and corresponding headways will not affect our analysis significantly as a result of the previously described filtering efforts.

Figure 4.6b shows the empirical headway distributions within each velocity range and their extreme values and centroids. It indicates that the mean values of headways decrease with the velocities and finally reach the saturation headway with respect to the high velocities. Kolmogorov-Smirnov (K-S) hypothesis test results show that all headway distributions follow a family of lognormal distributions.

Suppose that within the kth velocity group, the collected headway $h_i(t)$ follows a lognormal type velocity-dependent distribution Γ^k, i.e.,

$$h_i(t) \sim \text{Log-}\mathcal{N}\left(\mu_k, \sigma_k^2\right), \quad v_i(t) \in \mathbb{S}_k \qquad (4.20)$$

where μ_k is the location parameter and σ_k is the scale parameter of lognormal distributions.

The PDF of $h_i(t)$ can be written as

$$f(h_i(t)) = \frac{1}{\sqrt{2\pi}\sigma_k h_i(t)} \exp\left[-\frac{(\log h_i(t) - \mu_k)^2}{2\sigma_k^2}\right], \ h_i(t) > 0, \ v_i(t) \in \mathbb{S}_k$$
(4.21)

The CDF of $h_i(t)$ can be written as

$$\Gamma^k = \Phi\left(\frac{\log h_i(t) - \mu_k}{\sigma_k}\right), \ v_i(t) \in \mathbb{S}_k \qquad (4.22)$$

where Φ is the standard normal distribution function.

The $\alpha/2$ quantile of the lognormal distribution can be calculated by the inverse CDF of $h_i(t)$ as follows

$$\Gamma_{\alpha/2}^{-k} = \exp[\sigma_k\Phi^{-1}(\alpha/2) + \mu_k], \ v_i(t) \in \mathbb{S}_k \qquad (4.23)$$

where $\Gamma_{\alpha/2}^{-k}$ is the $\alpha/2$ quantile of Γ^k with respect to the kth velocity range, Φ^{-1} is the inverse function of standard normal distribution.

To discard abnormal patterns, we only focused on the headways within a confidence interval at the significance level of α as $[\Gamma_{\alpha/2}^{-k}, \Gamma_{1-\alpha/2}^{-k}]$, where $\Gamma_{\alpha/2}^{-k}$ and $\Gamma_{1-\alpha/2}^{-k}$ denote the inverse function of the CDF at $\alpha/2$ and $1 - \alpha/2$, respectively.

We estimated the parameters of lognormal distributions for each velocity range based on the filtered NGSIM trajectory data. Figure 4.7 shows the mean $\hat{\mu}_k$ and standard deviation $\hat{\sigma}_k$ of the headways' natural logarithm. From this, we determined the confidence intervals of headway $[\Gamma_{\alpha/2}^{-k}, \Gamma_{1-\alpha/2}^{-k}]$ at the significance level of $\alpha = 0.05$. The mean headway values indicate that the lower and upper bounds of headway decrease monotonically as velocity increases, which is in accordance with Fig. 4.6b.

Fig. 4.7 The estimated shape coefficients of lognormal distributions

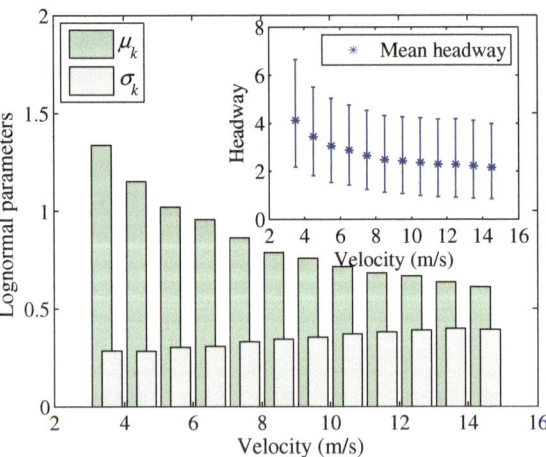

4.3.2.2 Tau Value Estimation Algorithm

We discretize $[\Gamma^{-k}_{\alpha/2}, \Gamma^{-k}_{1-\alpha/2}]$ into N subsets uniformly, each of which occupies a headway range of $[\Gamma^{-k}_{1-\alpha/2} - \Gamma^{-k}_{\alpha/2}]/N$. We will only study the effective headway adjustment behaviors where $h_j(t)$ belongs to the mth subset and $h_j(t+T)$ belongs to a different nth subset. This method filters out data that were collected during periods of zero action, allowing us to isolate and focus on headway adjustment data. Furthermore, we screened out data from headway changes caused by the ac/deceleration of leading vehicles. That is, $\left(v_j(t), h_j(t), h_j(t+T)\right)$ is a valid sample, if and only if the velocity change of the leading vehicle is less than the threshold of ϵ, i.e., $\left|v_{j-1}(t+T) - v_{j-1}(t)\right| \leqslant \epsilon$. We will discuss different velocity change thresholds (i.e., $\epsilon = 0.25$ m/s for $T = 0.5$ s; $\epsilon = 0.5$ m/s for $T = 1.0$ s; $\epsilon = 1.0$ m/s for $T = 2.0$ s and $T = 5.0$ s). As a result, $\mathbb{S}_k$ will be divided into four groups:

- if $h_j(t)$ and $h_j(t+T)$ are part of the same subset, the triple $\left(v_j(t), h_j(t), h_j(t+T)\right)$ is categorized as part of $\mathbb{S}^*_k$, which is the group that did not show significant adjustment behaviors;
- if $h_j(t) \to h_j(t+T)$ is an effective headway adjustment, and $h_j(t) > h_j(t+T)$, the triple is categorized as part of $\mathbb{S}^-_k$, which is the group that accelerated during headway adjustment;
- if $h_j(t) < h_j(t+T)$, the triple is categorized into the $\mathbb{S}^+_k$ group;
- if $v_{j-1}(t) \to v_{j-1}(t+T)$ is not an effective headway adjustment, the triple is categorized as part of the $\mathbb{S}^\Delta_k$ group, where velocity change of the leading vehicle is larger than ϵ.

Then, $\mathbb{S}^*_k \cup \mathbb{S}^-_k \cup \mathbb{S}^+_k \cup \mathbb{S}^\Delta_k = \mathbb{S}_k$, and $\mathbb{S}^*_k \cap \mathbb{S}^-_k \cap \mathbb{S}^+_k \cap \mathbb{S}^\Delta_k = \emptyset$. The numbers of triples within each subset are denoted as $\mathcal{J}^*_k$, $\mathcal{J}^-_k$, $\mathcal{J}^+_k$, $\mathcal{J}^\Delta_k$, satisfying $\mathcal{J}^*_k + \mathcal{J}^-_k + \mathcal{J}^+_k + \mathcal{J}^\Delta_k = \mathcal{J}_k$, $\forall k$. Thus, $\hat{T}_k$ and $\hat{p}_k$ are calculated as

$$\hat{T}^{\pm}_k = \frac{1}{\mathcal{J}^{\pm}_k} \sum_{j=1}^{\mathcal{J}^{\pm}_k} \frac{h_j(t)T}{h_j(t+T) - h_j(t)} \tag{4.24}$$

$$\hat{p}_k = \frac{\mathcal{J}^-_k}{\mathcal{J}^-_k + \mathcal{J}^+_k} \tag{4.25}$$

where $\hat{T}^-_k$ and $\hat{T}^+_k$ are the Tau factors for the kth velocity range, $\hat{p}_k$ is probability estimate for the kth velocity range.

4.3.2.3 The Influence of Sampling Time Interval

Choosing an appropriate sampling time interval is a difficult problem. Generally, the chosen T needs to be long enough to allow the driver to react. Therefore, it cannot be less than 0.5 s. A smaller sampling time interval will result in the inability

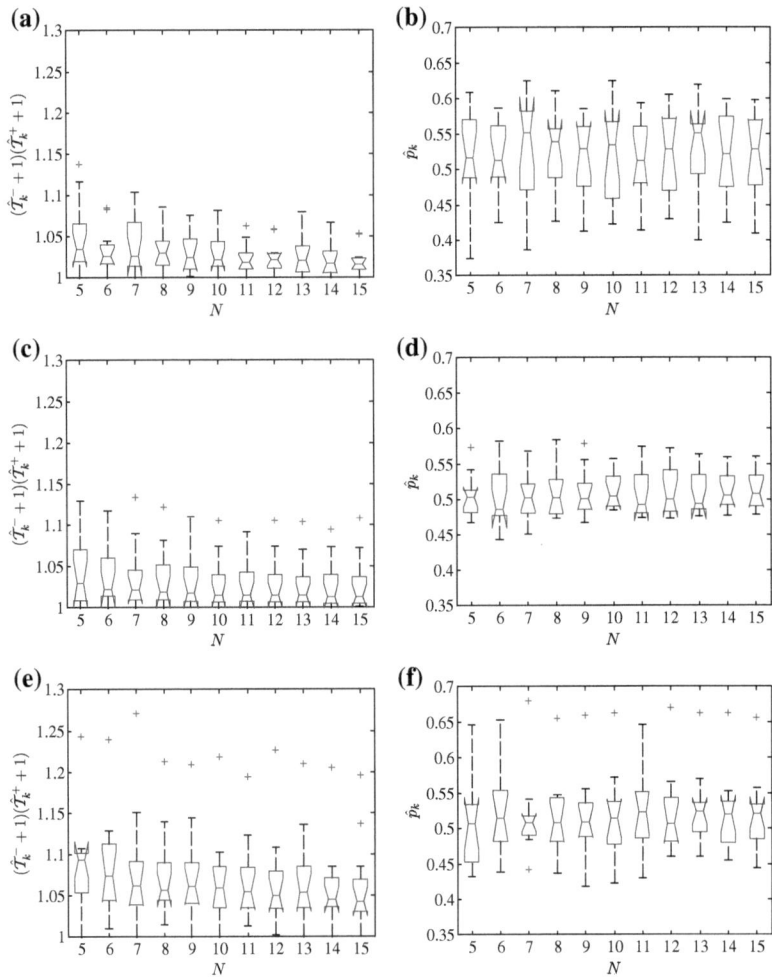

Fig. 4.8 Variation of the two behavioral statistics T_k^* and $\hat{p}_k$. **a** $T = 0.5s$, $F = 1.07$, $p = 0.40$. **b** $T = 0.5s$, $F = 0.08$, $p = 0.99$. **c** $T = 1.0s$, $F = 0.34$, $p = 0.97$. **d** $T = 1.0s$, $F = 0.19$, $p = 0.99$. **e** $T = 2.0s$, $F = 0.30$, $p = 0.98$ **f** $T = 2.0s$, $F = 0.13$, $p = 0.99$

to smooth out the measurement noise generated by vision-based vehicle trajectory retrieval, which may lead to incorrect conclusions. On the other hand, headway adjustment actions can be viewed as a series of temporally discrete ac/decelerations, where the driver rapidly makes one decision/action after another. If the sampling time interval is larger than 5.0 s, we may be unable to capture many of the driver's actions. We compare four sampling time intervals: 0.5, 1.0, 2.0 and 5.0 s based on the *mean criterion* and *variance criterion*. Figure 4.8 compares the variations of $(T/\hat{T}_k^- + 1)(T/\hat{T}_k^+ + 1)$ and $\hat{p}_k$, $k = 1, 2, ..., 12$, with respect to the subset number, N ($N = 5, 6, ..., 15$), and the sampling time interval T ($T = 0.5, 1.0, 2.0, 5.0$ s).

To further explain these results, we incorporated the ANOVA to determine whether or not the mean Tau values are equal for different headway groups (each value of N represents one division strategy). The p-value results from ANOVA depend on the assumption of independent and normal distributions with constant variances. More specifically, one-way/single-factor ANOVA is used to test the statistical significance by comparing the F test statistic. In ANOVA, large differences between the central lines of the boxes correspond to a small p-value derived from the large values of the F statistic. In addition, a small p-value usually indicates significant differences between column means. When there are only two means to compare, the t-test and the ANOVA F-test are equivalent; the relation between ANOVA and t-test is given by $F = t^2$.

As shown in Fig. 4.8, the means of $(T/\hat{T}_k^- + 1)(T/\hat{T}_k^+ + 1)$ indicate significant differences among some of the means in the N division strategies under $T = 0.5$ s. This is because the probability of the observed F being larger than the F statistic is relatively small at 0.40.

The p-values are close to 1.00 for all N division strategies at $T = 1.0$ s, $T = 2.0$ s and $T = 5.0$ s. This indicates that the means of $(T/\hat{T}_k^- + 1)(T/\hat{T}_k^+ + 1)$ for each group are equal. Since we do not know the exact distributions or a $priori$ knowledge of $\hat{T}_k^-$, $\hat{T}_k^+$ and $\hat{p}_k$, it is difficult to accurately conduct the hypothesis test to compare $(T/\hat{T}_k^- + 1)(T/\hat{T}_k^+ + 1)$ and 1.0, $\hat{p}_k$ and 0.5, respectively. Therefore, we will focus on the empirical statistical properties based on the field data.

It is interesting to note that the variation range of $(T/\hat{T}_k^- + 1)(T/\hat{T}_k^+ + 1)$ stays within $(1.0, 1.1)$ when $T \leqslant 1.0$ s; but significantly increases with T when $T \geqslant 2.0$ s. This may be caused by the omission of driver actions due to large T values. In short, based on the statistics of $(T/\hat{T}_k^- + 1)(T/\hat{T}_k^+ + 1)$ and $\hat{p}_k$, $T = 1.0$ s is the best sampling time interval. In addition, $\hat{p}_k$ for all N can be generally viewed as the same at $T = 0.5$ s, $T = 1.0$ s, $T = 2.0$ s and $T = 5.0$ s. This is because the p-values for all four cases are all close to 1.0. From the statistics in Fig. 4.8, the box-and-whisker diagram graphically depicts the lower quartile, median, and upper quartile. We find that when $T = 1.0$ s, the ranges between the upper and lower quartiles for both $(T/\hat{T}_k^- + 1)(T/\hat{T}_k^+ + 1)$ and $\hat{p}_k$ are the lowest. This empirical observation validates the chosen $T = 1.0$ s is more reasonable sampling time interval for an asymmetrical behavioral analysis.

4.3.2.4 The Influence of Division Strategies

Whether the division strategies (N) for headway change affect our tests on the Tau values also requires careful examination. For example, Table 4.4 shows the variations of the average probability for reaching a smaller headway $\overline{p}$ and the average asserting index $\overline{(T/\hat{T}^- + 1)(T/\hat{T}^+ + 1)}$ weighted by the number of samples, with respect to different N values (from 5 to 15). The sampling time interval is chosen as $T = 1.0$ s.

Table 4.4 The estimated parameters with respect to N, $T = 1.0\,\text{s}$, ($\sum_k \mathcal{J}_k = 56436$)

N	$\sum_k \mathcal{J}_k^-$	$\sum_k \mathcal{J}_k^+$	$\frac{\sum_k(\mathcal{J}_k^- + \mathcal{J}_k^+)}{\sum_k(\mathcal{J}_k - \mathcal{J}_k^\Delta)}$ (%)	$\overline{T^*}$	$\overline{\hat{p}}$
5	3859	3723	17.97	1.03	0.51
6	4590	4484	21.50	1.03	0.51
7	5401	5239	25.21	1.03	0.51
8	6268	6010	29.10	1.02	0.51
9	7025	6722	32.58	1.02	0.51
10	7614	7207	35.12	1.02	0.51
11	8226	7822	38.03	1.02	0.51
12	8842	8367	40.78	1.02	0.51
13	9260	8912	43.06	1.02	0.51
14	9984	9476	46.11	1.02	0.51
15	10379	9886	48.02	1.02	0.51

$$\overline{\hat{p}} = \frac{\sum_k [(\mathcal{J}_k^- + \mathcal{J}_k^+)\hat{p}_k]}{\sum_k (\mathcal{J}_k^- + \mathcal{J}_k^+)} = \frac{\sum_k \mathcal{J}_k^-}{\sum_k (\mathcal{J}_k^- + \mathcal{J}_k^+)} \tag{4.26}$$

$$\overline{T^*} = \overline{(T/\hat{T}^- + 1)(T/\hat{T}^+ + 1)} = \frac{\sum_k [(\mathcal{J}_k^- + \mathcal{J}_k^+)(T/\hat{T}_k^- + 1)(T/\hat{T}_k^+ + 1)]}{\sum_k (\mathcal{J}_k^- + \mathcal{J}_k^+)} \tag{4.27}$$

where $\overline{\hat{p}}$ is the average probability of headway decline, $\overline{T^*}$ is the weighted average of T^* statistics. Results indicate that $\overline{\hat{p}}$ maintains a relatively constant 0.51, and $\overline{T^*}$ is around 1 for all the headway division strategies.

Figure 4.9 presents the distributions of $\overline{T^*}$ and $\overline{\hat{p}}$ with respect to both the number N of headway division groups and the sampling time interval T under the same velocity change threshold of $\epsilon = 1.0\,\text{m/s}$. The obtained p-values are approximately 0.00, strongly indicating that the means of $\overline{T^*}$ and $\overline{\hat{p}}$ at different time sampling intervals are not the same. It is worthwhile to point out that the different means, with respect to various T, are significantly different. This is consistent with the results shown in Fig. 4.9, where we concluded the same means for $(T/\hat{T}_k^- + 1)(T/\hat{T}_k^+ + 1)$ and $\hat{p}_k$, in terms of different headway division strategies N (5, 6, ..., 15) at each T. The comparison indicates that the headway division strategies have no significant influences on the statistical tests; while the sampling time intervals will significantly impact statistical tests.

4.3.2.5 The Influence of Velocity Change Threshold

As mentioned in the third step of the data filtering, whether the velocity change threshold of the leading vehicle affect our tests on the Tau values is worthy for investigations. In the above discussions, the velocity change threshold is set $\epsilon = 1.0\,\text{m/s}$, below which changes in velocity of the leading vehicle are minor.

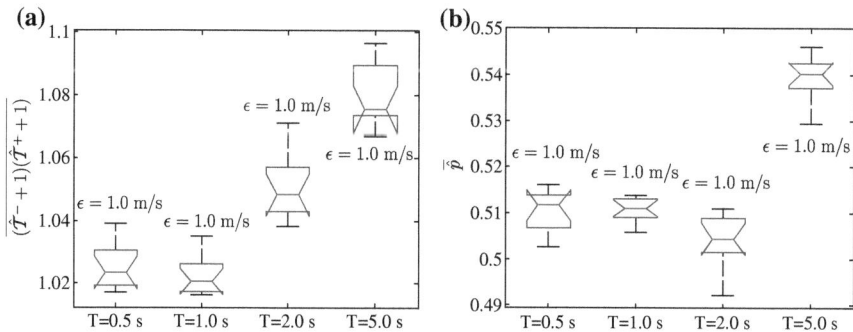

Fig. 4.9 Variation of the two behavioral statistics with respect to T: $\overline{T^*}$ and $\overline{p}$. **a** $F = 105.92$, $p = 0.00$. **b** $F = 120.99$, $p = 0.00$

If we estimate the approximate acceleration and deceleration rates for each sampling time interval T, e.g., the leading vehicle could brake up to $-\epsilon/T = -0.2$ m/s² (for $T = 5.0$ s) and $-\epsilon/T = -0.5$ m/s² (for $T = 2.0$ s), these minimal decelerations rates within one sampling time interval would be reasonable for the aforementioned assumption of the stable leading vehicle. However, the deceleration rates will be $-\epsilon/T = -1.0$ m/s² (for $T = 1.0$ s) and $-\epsilon/T = -2.0$ m/s² (for $T = 0.5$ s), they may infer to solid braking maneuvers, and therefore the car-following behaviors are not stable.

In order to solve this problem, we discuss a minimal deceleration rate of $-\epsilon/T = -0.5$ m/s² (or maximal acceleration rate of $-\epsilon/T = 0.5$ m/s²) for the leading vehicle in this subsection. We only choose smaller velocity change thresholds for $T = 0.5$ s and $T = 1.0$ s, i.e., $\epsilon = 0.25$ m/s and $\epsilon = 0.5$ m/s, respectively. We filter the samples in steady car-following states because the acceptable velocity change thresholds become smaller. Recall Table 4.3, for $T = 0.5$ s, the sample size reduces from 13,827 to 7,503 (46 % reduction); for $T = 1.0$ s, the sample size reduces from 42,199 to 30,376 (28 % reduction).

Though the remaining sample sizes decrease to a great extent, as shown in Fig. 4.10, the two behavioral statistics $\overline{T^*}$ and $\overline{p}$ do not perform differently with respect to both the number N of headway division groups and the sampling time interval T under the various velocity change thresholds. Using the *mean and variance criteria* to check the best sampling time interval out of the four alternatives, we can see that the interquartile ranges of $T = 1.0$ s is also the smallest. The corresponding mean values are $\overline{T^*} = 1.02$ and $\overline{p} = 0.51$, which are almost the same as Table 4.4 and Fig. 4.10. This consistency indicates that the velocity change threshold has no significant influence on the Tau Theory statistics.

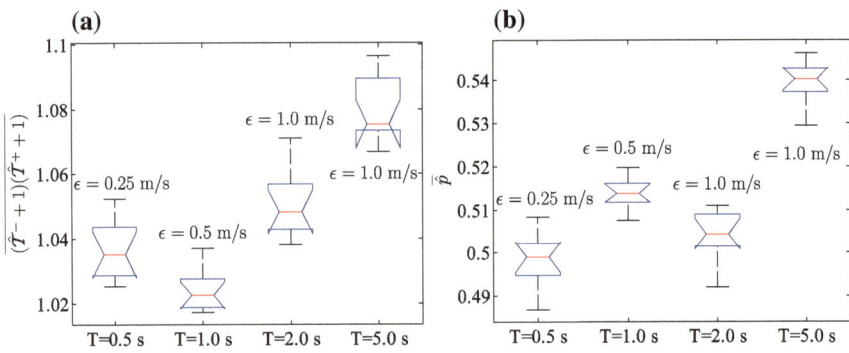

Fig. 4.10 Variations of the two behavioral statistics with respect to T (various ϵ): $\overline{T^*}$ and $\overline{\hat{p}}$. **a** $F = 77.99$, $p = 0.00$. **b** $F = 124.37$, $p = 0.00$

4.3.2.6 Testing Results on Tau Values

Figure 4.11 shows the mean values, i.e., the 25th and 75th percentiles of $\hat{T}_k^-$ and $\hat{T}_k^+$ within the kth velocity range with a sampling time interval of $T = 1.0$ s. We divided the headway confidence interval into 10 subgroups, i.e., $N = 10$. The mean values of $\hat{T}_k^-$ and $\hat{T}_k^+$ decreased as the velocity increase until $v > 15$ m/s. This was mainly because in the NGSIM dataset, we did not have enough free flow headway samples where $v > 15$ m/s.

Table 4.5 shows the number of samples based on the division strategies of valid triples $\left(v_j(t), h_j(t), h_j(t + T)\right)$, where $\mathcal{J}_k^* + \mathcal{J}_k^- + \mathcal{J}_k^+ + \mathcal{J}_k^\Delta = \mathcal{J}_k$ is satisfied. The percentages of the pruned data, as a result of the leading vehicle velocity change, are between 24 to 30 %, and shown as $\mathcal{J}_k^\Delta / \mathcal{J}_k$. If the velocity change of the leading vehicle is larger than ϵ, we regard the headway adjustment may be caused by the

Fig. 4.11 The estimated Tau values

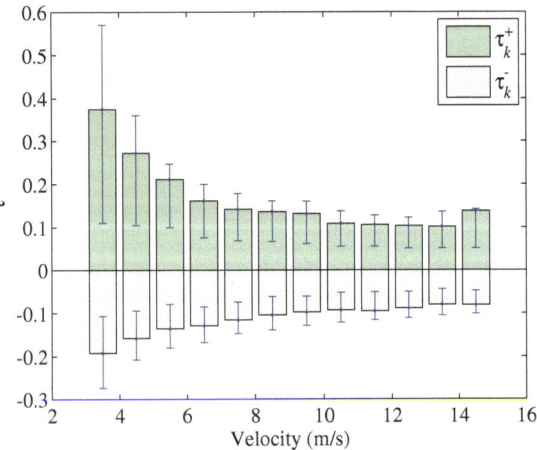

Table 4.5 The parameters estimated from the NGSIM dataset ($T = 1.0$, $N = 10$)

k	v	$\mathcal{J}_k$	$\mathcal{J}_k^\Delta$	$\frac{\mathcal{J}_k^\Delta}{\mathcal{J}_k}$ (%)	$\mathcal{J}_k^-$	$\hat{T}_k^-$	$\mathcal{J}_k^+$	$\hat{T}_k^+$	T_k^*	$\hat{p}_k$	$\frac{\mathcal{J}_k^+ + \mathcal{J}_k^-}{\mathcal{J}_k - \mathcal{J}_k^\Delta}$ (%)
1	3–4	1113	330	30	249	−0.19	258	0.37	1.11	0.49	65
2	4–5	1924	518	27	446	−0.16	424	0.27	1.07	0.51	62
3	5–6	1545	435	28	319	−0.14	332	0.21	1.05	0.49	59
4	6–7	3308	924	28	671	−0.13	548	0.16	1.01	0.55	51
5	7–8	5103	1354	27	875	−0.12	695	0.14	1.01	0.56	42
6	8–9	3705	893	24	520	−0.10	551	0.14	1.02	0.49	38
7	9–10	7296	1755	24	1030	−0.10	907	0.13	1.02	0.53	35
8	10–11	9904	2485	25	1142	−0.09	1107	0.11	1.01	0.51	30
9	11–12	5047	1206	24	571	−0.10	569	0.11	1.00	0.50	30
10	12–13	8028	2062	26	806	−0.09	841	0.10	1.01	0.49	28
11	13–14	6979	1684	24	699	−0.08	725	0.10	1.01	0.49	27
12	14–15	2529	636	25	286	−0.08	250	0.14	1.05	0.53	28

nonstationary behaviors of the leading vehicle but not the following driver's nature stochastic behavior in such a case. Thus, we eliminate these samples with an obvious velocity change of the leading vehicle.

The sampling percentages of headway change or effective headway adjustment are between 27 and 65 %. We find this percentage decreases when velocity increases, headway tends to be more stable for the higher velocity ranges. This may be caused by relatively more random maneuvers perform in the low-velocity state. It is equivalent to simplify the assertion of $(T/\hat{T}_k^- + 1)(T/T_k^+ + 1)$ as $T + T_k^+ + T_k^- = 0$. It is estimated by substituting the mean values of $\hat{T}_k^-$ and $\hat{T}_k^+$. It is shown that $T + \hat{T}_k^- + \hat{T}_k^+$ concentrates around zero with an acceptable accuracy given the Tau values.

The relationship between Tables 4.4 and 4.5 is as follows: Table 4.4 summarizes the estimated $\overline{T}^*$ and $\overline{p}$ by checking the samples across all velocity ranges; Table 4.5 is designed to estimate $\hat{T}^*$ and $\hat{p}_k$ for a specific headway division $N = 10$. Based on these results, it is clear that the probability of reaching a smaller or larger headway keeps an acceptably constant of 0.5. This indicates a relatively constant tendency p to reach a headway that is smaller than preferred in different velocities.

The assertion, $(T/\hat{T}_k^- + 1)(T/\hat{T}_k^+ + 1) - 1$ can be empirically validated based on Table 4.5, since the values of $\hat{T}^*$ are between 1.0004 and 1.1103 for all the velocity range. When velocities are higher than 5 m/s, the values of $\hat{T}^*$ are between 1.00 and 1.05, which are almost equal to the previous assertion. This means that the reciprocal relationship between T^- and T^+ holds, especially when there is less traffic congestion (e.g. velocities are higher than 5 m/s). The agreement between the predictions of this extended Tau Theory and the empirical observations indicates that the physiological Tau characteristics can be viewed as a dominant factor that implicitly governs driving behavior.

As shown in Fig. 4.11, $|\hat{T}^-|$ and $\hat{T}^+$ roughly decrease as velocity increases. This can be explained as drivers being more careful at higher velocities. It is interesting to note that the percentage of headway state transitions in all the effective samples, i.e., $(\mathcal{J}_k^+ + \mathcal{J}_k^-)/(\mathcal{J}_k - \mathcal{J}_k^\Delta)$ decreases with velocity. It may be because it is more difficult for a driver to maintain a preferred headway during congestion. Besides, the deviation of $\hat{T}^*$ from the asserted value of 1.0 implies that, at low velocities, it is increasingly difficult to describe human behavior using any single theory.

4.3.3 Discussions

In this section, we propose an extended Tau Theory model that assumes different T coefficients to maintain a constant headway in asymmetric car-following behaviors. This study uses multistage data filtering procedures to choose the steady car-following samples for each velocity range. Tests using the real-world data show that the predicted consistency of T coefficients agrees with the empirical observations. Although car-following behaviors can be better described using more complex microscopic models, e.g., the models listed in Brackstone and McDonald (1999), this finding reveals that the physiological Tau characteristics implicitly affect human driving behavior and the resulting traffic dynamics.

In summary, field tests show two contributions in this study:

1. In headway adjustments, divers have a relatively constant tendency p (close to 0.5) to reach a smaller headway than preferred at different velocities and relatively constant T^- and T^+ at different velocities, satisfying $(T/T^-+1)(T/T^++1) = 1$.
2. The influences of sampling time interval on the statistical features of the asymmetric stochastic Tau Theory-based car-following behaviors are significant ($T = 1.0\,$s outperforms other sampling time interval according to the mean and variance criteria), but the influences of headway division strategies and velocity change thresholds of leading vehicles are not prominent.

It is also important to note that there are still several areas for future research, such as:

(1) The above findings are mainly based on NGSIM trajectory datasets, in which the currently four available datasets (US Highway 101 dataset, I80 dataset, urban arterials Lankershim Boulevard dataset, and Peachtree dataset) were all collected during rush hours without enough higher speeds data. So we are still unable to examine the driving behaviors in high speed traffic flow. And the proposed Tau theory-based model may not be fit for free driving behaviors.

On the other side, the main concern in this study is to address the close following actions of drivers when they are trying to keep safety distances; while in high speed traffic flow, the interactions between vehicles are relatively weak and the dominant task of a driver becomes keeping the highest possible speed in such traffic scenarios. Therefore, we infer such driving actions should be described by another set of model rather than Tau theory. We would like to collect more data to validate our inference in the future.

(2) Our research only used headway records collected from a single driving scenario. As proven in previous research, drivers are inconsistent in their headway choice across different driving scenarios, e.g., Brackstone et al. (2009) concluded that the road type (motorway versus urban dual carriageway) did not seem to affect headway choice and the type of leading vehicle (truck/vans or cars) did influence the following headway; Michael et al. (2000) implemented an intervention to decrease tailgating during normal traffic flow among a large number of drivers on urban roadways; and Broughton et al. (2007) illustrated drivers' velocity variability increased under conditions of reduced visibility. However, we believe that their headway adjustment processes will still be governed by the extended Tau theory, even though the observed T may vary with respect to certain other factors, including heterogeneity of different kinds of drivers (aggressive versus passive), different vehicle types (cars versus trucks, buses), inaccurate headway perceptions in foggy weather, and lane changing/merging actions. We will collect more data to validate our conjectures.

(3) Another question is how far the T coefficients for an individual driver can deviate from the common values in a population. In NGSIM datasets, we monitored one only 550-meter long road segment. As a result, we were unable to collect driver action data for a long enough time period, or develop effective statistical tests on T coefficients for individual drivers.

To solve this problem, we can use laser or radar sensors to accurately and diligently measure the relative speed and the space gap between vehicles. With the aid of inertial navigation systems (INS) and global positioning systems (GPS), we can also measure the positions of vehicles (Li and Wang 2007; Skog and Handel 2009) at all times. These emerging technologies and fast developing in-vehicle equipment can help us collect long term leading vehicle tracking data on individual drivers that cannot be obtained via video-based traffic monitoring methods.

Chapter 5
Stochastic Fundamental Diagram Based on Headway/Spacing Distributions

5.1 Introduction

Traffic flows on highways are often described in terms of three variables: mean velocity v, traffic flow rate q, and traffic density ρ. Based on the definitions of these variables, we have $q = v\rho$. So we only need to establish the relations between either two variables. All the discussions below focus on congested flow, because q increases almost linearly with ρ in free flow.

Among various approaches in literature, the FD is the most widely used model to describe the equilibrium flow-density relationship. The efforts on developing traffic flow models to capture the FD can be traced back to Greenshields' work in early 1930s (Greenshields 1935). Different functions for modeling FD had been proposed in the last 75 years, including Greenshields' model, Greensburg's model, Underwood's model, and many more complex variations (Del Castillo and Benitez 1995a, b; Gilchrist and Hall 1989; Hall et al. 1986; Li 2008; Wu 2002; Wu et al. 2011). Recently, Del Castillo (2001) proposed three models for the concave flow-density relationship that was validated by freeway data and urban data.

Two associated questions on the FD have been carefully discussed during the last two decades. The first one is how to estimate the FD in practice, since the FD is usually defined as a function of the homogeneous equilibrium states which may not be exactly measured from empirical data. As a result, the measured mean values of the three variables over a long time interval are often used to approximately estimate the FD (Banks 1989, 2002). Traditionally, the data used to estimate the FD are obtained from loop detectors at a specific location and are aggregated over time intervals that range from 30 s to 5 min. Recently, vehicle trajectory data based FD estimation methods received increasing interests, since equilibrium and transient phases are difficult to distinguish at a specific location.

The second question is whether the wide scattering feature of flow-density plot is in the concordance with the FD assumption. For example, Kerner (2001, 2009, 2004) argued that FD should not be used, because it failed to well account for the scattering features of flow-density plot. Differently, Treiber et al. (2006b) thought that the

© Tsinghua University Press, Beijing and Springer-Verlag Berlin Heidelberg 2015
X. (M.) Chen et al., *Stochastic Evolutions of Dynamic Traffic Flow*,
DOI 10.1007/978-3-662-44572-3_5

wide scattering of flow-density data could be reproduced by taking into account the variation of propagation speed or the local velocity variance caused by the variation of the netto time gaps among successive vehicles, and the delayed adaptation of the driving behaviors (Nishinari et al. 2003; Treiber and Helbing 2003; Treiber et al. 2006b).

In some recent approaches, these two questions were studied in a united way. The FD was linked to a certain car-following model so that we can combine the information provided by vehicle trajectory data and loop detector data. Particularly, Newell's simplified car-following model attracts increasing attentions (Newell 2002). Theoretical analysis and field data validation show that such a simple model provides an acceptable description of car-following dynamics (Ahn et al. 2004; Chiabaut et al. 2009, 2010; Kim and Zhang 2008; Lu and Skabardonis 2007; Wang and Coifman 2008; Yeo 2008; Yeo and Skabardonis 2009). As predicted by Newell (2002), it provides a starting point for investigating more complex traffic phenomena. As shown in Chiabaut et al. (2009), the assumption that FD holds in stable platoons is equivalent to the assumption that vehicles within the platoon follow Newell's simplified car-following model. How to obtain the FD based on the simplified model was then presented in Chiabaut et al. (2009, 2010), Duret et al. (2008), Yeo (2008). Kim and Zhang (2008) further explained the scattering feature in terms of random fluctuations of gap time based on the simplified car-following model. It provides an analytical framework to study how a perturbation of car-following spacing may grow and influence the flow-density plot. This approach is in consistent with Forbes (1963) and it indicates that we may discuss congested flows based on Newell's simplified car-following model. However, the region covered by the scattering points in the flow-density plot was not further estimated. More recently, Li and Zhang (2011) assumed that the congestion branch of the FD consisted of several velocity-constant fluctuations that meant the state transitions in the neighborhood of straight lines emanating from the origin of the flow-density plane. Li et al. (2011) analyzed how reliable the LWR model prediction would be when the uncertainties were introduced into the FD.

Following these approaches, we aim to establish a tight link between the spacing/headway distribution model and the wide scattering feature of flow-density plot based on the aforementioned discussions in this chapter. The motivations are trifold:

(1) We would like to provide a more rigorous relation between the distributions of microscopic-level traffic measurements (headways/spacings) and macroscopic-level traffic measurements (flow rate/density). There are lots of other models dedicated for individual driving behavior and the FD. For example, the linkage between GHR car-following model and the FD (Gazis et al. 1961). But, not many of them emphasized on the distribution features of measurements on two levels. Several models had been attempted to in this direction. However, few of them had clearly simultaneously described the distribution models on two levels and specified the model parameters. Moreover, no previous models had given explicit formulas on the mapping relation between distributions between two

levels. So, our model is an initial attempt to account for the implicit links and the distribution of measurements on two levels;

(2) The origins of wide scattering features of the points in the density-flow plot had attracted consistent interests (e.g., Chen et al. 2010b; Li et al. 2011). We aim to provide a simple and intuitive explanation that can fit with both microscopic-level and macroscopic-level traffic measurements, and meanwhile keeping limited possible assumptions in our proposed model;

(3) The model proposed is a good starting point to analyze several other traffic phenomena. One important application is to measure the FD curve similar to Chiabaut et al. (2009, 2010).

In this chapter, first, we review previous models on the scattering features and examine the stochastic description of drivers' tendency on spacing/headway choices. To distinguish the influences of aggregation time interval on the joint distribution of flow-density, we categorize the sampled parameters for congested traffic flow into two kinds: one kind of data are sampled from homogeneous platoons during short enough time periods; and the rest of the data are sampled from heterogeneous vehicle platoons during longer time periods. Then, we discuss the statistical relations between the microscopic average headway distribution in a homogeneous platoon and the resulting macroscopic flow-density plot; and an algorithm is given to project scattering features from the headway distributions with respect to different velocity ranges. Field tests on California PeMS data verify the hypothetical similarity between the estimated and empirical distributions of flow/density, and then prove the feasibility of the proposed stochastic platoon model. We will further discuss the average headway distribution in a heterogeneous platoon and demonstrate how to estimate its corresponding probabilistic boundaries in flow-density plot that characterizes the scattering features.

5.2 Newell's Simplified Model and Its Stochastic Extension

First, we will briefly revisit Newell's simplified car-following model. In short, this model assumes that if the vehicles are running on a homogeneous highway, the time-space trajectory of the following vehicle is essentially the same as the leading vehicle, except for a transition in both space and time; (see Fig. 5.1a). Indeed, we adopt and extend Newell's simplified model for two reasons:

(1) We aim to find a simple yet effective way to decouple the interlaced dependence relationship between velocity, density, and headway. When we introduce the piecewise linear trajectory, we implicitly presume a constant velocity within each linear segment and then we can handle the variation of spacing and headway in a much easier mode;

(2) Newell's simplified model is an accurate enough abstraction of the reality for our study.

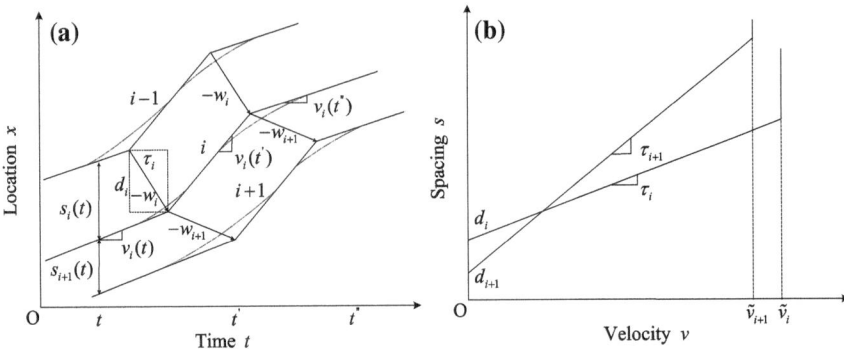

Fig. 5.1 Piecewise linear approximation to vehicle trajectories in Newell's model. **a** Vehicle trajectories; **b** Relation between spacing and velocity

It is supposed that the $(i - 1)$th vehicle is ahead of the ith vehicle, and their velocities are $v_{i-1}(t)$ and $v_i(t)$, respectively. When the leading $i - 1$th vehicle alters its velocity from v to v', it acquires a nearly constant velocity for some time. The following ith vehicle will adjust its velocity in the same way after a space displacement of d_i and an adjusting time of τ_i to reach the preferred spacing for a new velocity. For simplicity, Newell (2002) assumed $v_{i-1}(t) = v_i(t) = v$ and the continuous trajectory evolution (the dashed curve in Fig. 5.1a) can be approximated by piecewise linear curves (the solid line segments in Fig. 5.1a). These linear relationships between spacing and velocity are written as

$$s_i(t) = d_i + v_i(t)\tau_i, \ 0 \leqslant v_i(t) \leqslant \tilde{v}_i \tag{5.1}$$

where the spacing $s_i(t)$ refers to the distance between the front of the $i - 1$th vehicle and the front of the ith vehicle at time t, $\tilde{v}_i$ is the allowable velocity of the ith vehicle. As shown in Fig. 5.1b, these relationships are depicted as piecewise linear lines in the spacing-velocity plot. $s_i(t)$ is the vertical axis intercept, τ_i is the slope. We assume that d_i and τ_i are the inherent properties of the ith vehicle, and not relevant to $v_i(t)$. $\{d_1, d_2, \ldots, d_n\}$, and $\{\tau_1, \tau_2, \ldots, \tau_n\}$ of a platoon are i.i.d. random variables.

In short, the kernel of Newell's simplified model (Newell 2002) lies in the following assumption:

Definition 5.1 The *Newell Condition* is that two arbitrarily consecutive vehicles in a platoon travel on a homogeneous road essentially maintain the same time-space trajectory shape expect for a translation in space and in time.

Some efforts were made to develop methods for the estimation of wave velocities between several successive detectors, e.g., oblique cumulative curves (Cassidy and Bertini 1999), cross correlation method (Coifman et al. 2003). Careful investigations of NGSIM trajectory data (NGSIM 2006) show that the time-space trajectory of the following vehicle is not exactly the same as the leading vehicle, even when the transition in both space and time are considered. Newell (1965) used two separated curves in the spacing-velocity plot to depict the difference of acceleration and deceleration behaviors in congested traffic and pointed out that the car-following instability would lead to a growth of a small perturbation.

Yeo and Skabardonis (2009) followed this idea and interpreted the stop-and-go traffic phenomenon by an asymmetric microscopic driving behavior theory. They assumed that the deceleration curve (D-curve) was the lower bound or the minimum spacing preventing front-end collision even in the worst case of the leading vehicle's emergency braking. On the other hand, the acceleration curve (A-curve) is the upper bound of spacing that ensures a safe driving spacing even in the case of the following vehicle's accelerating and the leading vehicle's sudden stop. In our Chap. 4, the asymmetric car-following behaviors were explained by using the stochastic extension of the physiological Tau Theory.

The shape of these two curves depends on the maximum deceleration rate $a_{\max,\, i-1}$ of the leading vehicle, the comfortable deceleration rate b_i of the following vehicle, the ac/decelerating reaction time τ_i^A and τ_i^D, satisfying $\tau_i^A > \tau_i^D$. Thus, Yeo (2008) modeled these two curves as

$$
\begin{cases}
\text{A-curve: } s_i^A(t) = v_i(t)\tau_i^A + (a_{\max,\, i-1}^{-1} - b_i^{-1})v_i^2(t)/2 + s_{\text{stop}}^A \\
\text{D-curve: } s_i^D(t) = v_i(t)\tau_i^D + (a_{\max,\, i-1}^{-1} - b_i^{-1})v_i^2(t)/2 + s_{\text{stop}}^D
\end{cases}
\tag{5.2}
$$

where s_{stop}^A is the jam spacing in acceleration phase, s_{stop}^D is the jam spacing, satisfying $s_{\text{stop}}^A > s_{\text{stop}}^D$, A and D indicate acceleration and deceleration, respectively.

As shown in Yeo (2008), the A/D curves enclosed a 2D region in the spacing-velocity plot (see Fig. 5.2a, b). The coasting behavior around the A/D curves implied that the equilibrium traffic flow state lied on this 2D region. Thus, the A/D curves corresponded with the boundaries of congested flow points in flow-density plot.

Differently, Del Castillo (2001) studied the transition of traffic states within this 2D region. He developed an analytical approximation for the normalization probability of the growth/decay of a perturbation based on the headway distribution and on the ratio of the wave speed to the maximum speed. Kim and Zhang (2008) showed that the growth/decay of perturbations caused by random fluctuations of gap time would lead to a random transition of traffic state in this 2D region (see Fig. 5.2d). The stochastic wave propagation speeds are associated to the relative differences between gap times and the reaction times of drivers. Generally, such random state transitions are restricted in a 2D region in flow-density plot, whose boundaries encompassed the traffic wave structures. However, Kim and Zhang (2008) did not address the empirical statistics of the spacings/headways and thus did not derive the boundaries of this 2D region.

Let us take the trajectories shown in Fig. 5.2c as an example. Here, the waves of acceleration and deceleration perturbations occur in the order of $w_1 \rightarrow w_2 \rightarrow w_3$. Suppose P_1, P_2, P_3 are the corresponding instantaneous transition states of Platoon 1. As illustrated in Fig. 5.2d, the transition from P_1 to P_3 yields a random walk in the flow-density plot, including two state transitions. The first one satisfies $\overrightarrow{P_1P_2} \propto -\overrightarrow{w}_1, v_1 \rightarrow v_2$, the second one satisfies $\overrightarrow{P_2P_3} \propto -\overrightarrow{w}_2, v_2 \rightarrow v_1$. Generally,

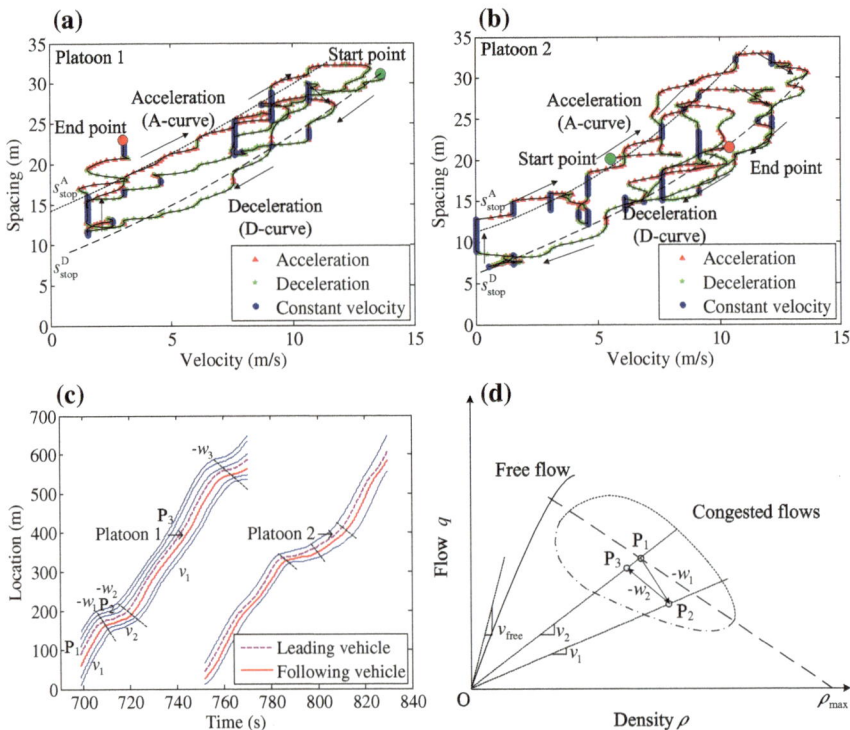

Fig. 5.2 Asymmetric characteristics of typical platoons in car-following processes. **a, b** Velocity spacing transitions observed in Highway 101 NGSIM dataset; **c** the platoon trajectories corresponding to **a** and **b**, respectively; **d** an illustration of the generated random walk in the flow-density plot for Platoon 1

this random walk is restricted in a 2D region in the flow-density plot, whose boundary encompasses the traffic wave structures.

One remaining problem is how to account for the observed spacing-velocity pairs outside the 2D region enclosed by A/D curves, since A/D curves were assumed to the maximum/minimum spacing in Yeo (2008). Chen et al. (2010b) solved this problem by providing a statistical explanation as follows. Noticing that drivers make random choices of spacing but with potential preferences, Chen et al. (2010b) studied the joint distribution of the spacing-velocity pairs and also the α-truncated conditional PDFs of spacing under different velocities. The velocity-dependent upper bounds are denoted as $F_{|v,\,1-\alpha/2}^{-1}$ and the lower bounds are $F_{|v,\,\alpha/2}^{-1}$, $v \in (0,\ v_{\text{free}})$ at the same confidence level α. Here, $F_{|v}$ denotes the CDF conditional on velocity and $F_{|v}^{-1}$ means the inverse function. The α confidence level is introduced here to filter out the influences of some abnormal data. The spacing-velocity samples falling into the truncated PDF ranges may reflect transient driving actions. In other words, Chen et al. (2010b) defined the maximum/minimum normal spacing as the α-level statistical upper/lower bounds of spacings with respect to velocity.

Fig. 5.3 An illustration of the α-truncated marginal distributions of spacing

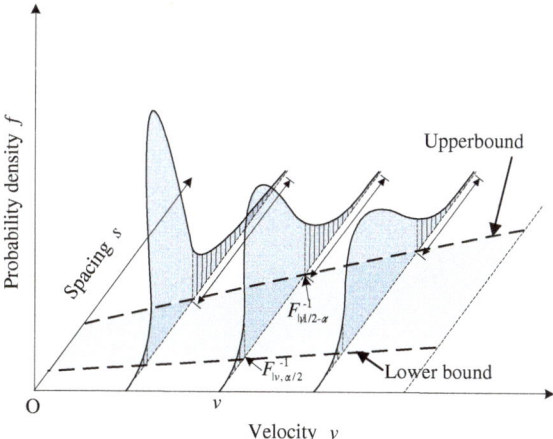

As shown in Fig. 5.3, by linking the lower and upper truncating points of each conditional PDFs of spacings under each range of velocity, the obtained joint PDF of spacing-velocity envelops a 2D region in spacing-velocity plot. Empirical data indicate that these two curves, i.e., upper and lower boundaries, are approximately two straight lines. Following the approach of Kim and Zhang (2008), Chen et al. (2010a) defined the maximal/minimal wave travel speeds w^{A} and w^{D} as

$$w^{A} = \frac{s_{\text{upper}}^{A}\tilde{v}^{+} - s_{\text{lower}}^{A}\tilde{v}^{-}}{s_{\text{upper}}^{A} - s_{\text{lower}}^{A}}, \quad w^{D} = \frac{s_{\text{upper}}^{D}\tilde{v}^{+} - s_{\text{lower}}^{D}\tilde{v}^{-}}{s_{\text{upper}}^{D} - s_{\text{lower}}^{D}} \tag{5.3}$$

where s_{upper}^{A} and s_{upper}^{D} are the minimum and maximum spacings after truncation at $\tilde{v}^{+}$, respectively; s_{lower}^{A} and s_{lower}^{D} are the minimum and maximum spacings after truncation at $\tilde{v}^{-}$, respectively (Fig. 5.4b).

Figure 5.4a illustrates the growth/decay of perturbations in flow-density plot; see also Kim and Zhang (2008). Figure 5.4b shows the relations between w^{A}, w^{D}, and the estimated (dashed) 2D region of congested flows. Chen et al. (2010a) further assumed that the transition of congested flows would not cross the corresponding boundaries determined by w^{A} and w^{D}. More precisely, the points of congested flow locate within the intersection of two sectors, i.e., the shaded area in flow-density plot: one sector centers at the original point and the slopes of its bounds are the minimal velocity $\tilde{v}^{-}$ and the maximum velocity $\tilde{v}^{+}$; the other sector centers at the point $(\rho_{\max}, 0)$ and the slopes of its bounds are the acceleration wave travel speed w^{A} and the deceleration wave travel speed w^{D}.

Figure 5.4c shows the random transition of the state P, characterized by $(\tau_{P}, -w_{P})$ in the spacing-velocity plot corresponding to Fig. 5.4a. Clearly, w^{A} and w^{D} directly correspond to the interceptions of the upper/lower boundaries of the 2D region in the flow-density plot. Figure 5.4d shows the stochastic property of an arbitrary vehicle i characterized by (τ_{i}, w_{i}) within the envelope curves of acceleration and deceleration

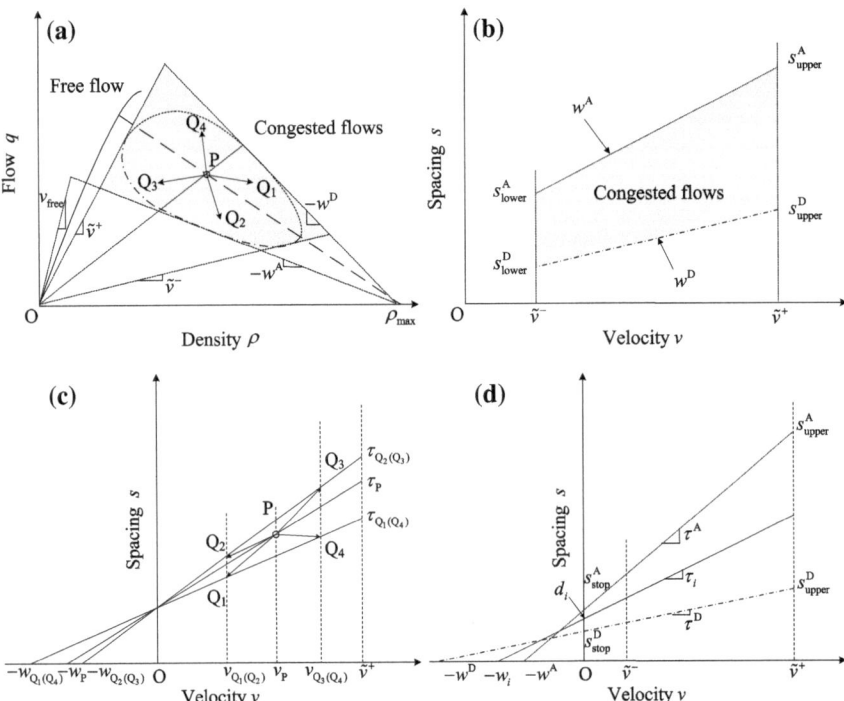

Fig. 5.4 Stochastic characteristics of congested traffic flow in FD. **a** The growth/decay of perturbations that is reflected in the flow-density plot. The transitions P → Q_1, P → Q_2, P → Q_3, and
P → Q_4 stand for deceleration disturbance vanishes, deceleration disturbance grows, acceleration
disturbance vanishes and acceleration disturbance grows, respectively. Detailed discussions can be
found in Kim and Zhang (2008); **b** the relation between w^A, w^D, and the estimated (*dashed*) 2D
region of the congested flows. Detailed discussions can be found in Chen et al. (2010b); **c** the random
state transitions shown in spacing-velocity plot; **d** the maximum/minimum stochastic propagation
wave speeds in spacing-velocity plot

characterized by (τ^A, w^A s_{stop}^A) and (τ^D, w^D s_{stop}^D), that satisfy $\tau^D < \tau_i < \tau^A$,
$w^A < w_i < w^D$ and $s_{\text{stop}}^D < d_i < s_{\text{stop}}^A$.

Although the model proposed in Chen et al. (2010a) recovered the implicit
links between the spacing-velocity distribution models and the flow-density plot,
it neglected one important fact: the flow-density plots are usually obtained by aggre-
gating the measured flow rate and density over one to several minutes. In other
words, the flow-density plots depict the traffic parameters over a vehicle platoon that
contains several vehicles. On the contrary, the spacing distributions are generally
measured between two consecutive vehicles. Therefore, we need to discuss how to
combine these two measurements collected on different time and space scale, rather
than simply map one into the other as Chen et al. (2010a). Besides, the distribution of
points in flow-density plot also needs further discussions. Surprisingly, we find that

such tendency can be implicitly determined by the spacing/headway distributions. We will further explain our findings in the following sections.

5.3 The Homogeneous Platoon Model

5.3.1 Basic Idea

A straightforward idea to connect the microscopic measurements of traffic flow (spacing/headway) and the macroscopic measurements of traffic flow (density/flow rate) are *deterministic reciprocal relations*:

$$q = 3600/h, \quad \rho = 1000/s \tag{5.4}$$

where q, ρ, s, and h are conventionally assumed to be deterministic measurements (usually the mean values) of flow rate (veh/h), density (veh/km), spacing (meters), and headway (seconds).

These fundamental formulas hold for homogeneous congested traffic flow and set up a good starting point for our further discussions. However, empirical traffic flows are not homogeneous due to various influencing factors, e.g., the mixture of various kinds of vehicles, heterogeneous driving behaviors, stochastic headway/spacing perception, etc.

From the viewpoint of platoon dynamics, we further categorize empirical congested traffic flows into two kinds in this chapter:

(1) Traffic flow is in a near homogeneous state. That is, the velocities are roughly the same in these so called homogeneous platoons. The spacings/headways between any two consecutive vehicles may be different, because empirical spacings/headways follow certain lognormal type distributions under a given velocity (Chen et al. 2010a, b; Jin et al. 2009). If we use loop detectors to collect traffic flow information, a sequence of vehicles recorded within a short time interval can be roughly viewed as belonging to a homogeneous platoon. In this chapter, we assume that when the aggregation time interval is set as $T = 30$, the sampled data correspond to homogeneous platoons.

(2) Traffic flow is far from homogeneous and can be depicted as heterogeneous. In other words, both the velocities and spacings/headways of a series of vehicles are different in these so called heterogeneous platoons. Usually, a large heterogeneous vehicle platoon consists of several small homogeneous vehicle platoons. If we use loop detectors to collect traffic flow information, the increase of the aggregation time interval often leads to an increasing loss of information and makes it difficult to build a simple distribution model for flow rate and density. In this chapter, we assume that when the aggregation time interval is set as $T \geqslant 5$ min, the sampled data often correspond to heterogeneous platoons.

For the first kind of traffic flows, we extend the deterministic reciprocal relations into the following stochastic reciprocal relations:

Proposition 5.1 *When velocity is given, the reciprocal of average headway of vehicles in a homogeneous platoon and the corresponding flow rate follow the same distribution; similarly, the reciprocal of the average spacing of vehicles in a homogeneous platoon and the corresponding density follow the same distribution.*

More precisely, we have the following relations in probability distribution

$$q \Leftrightarrow 3600/h, \quad \rho \Leftrightarrow 1000/s \tag{5.5}$$

We could map the distribution of average headways/spacings under different velocities to the distributions of flow rates/density in flow-density plot.

As shown in Fig. 5.5, it indeed provides a new interpretation on the distribution of points in flow-density plot. Different from the conventional orthogonal coordinates of density and flow rate, these points are now investigated under the polar coordinates of velocity and flow rate. Though the velocity in the first triple (velocity, headway, spacing) refers to individual velocity whereas the second triple (speed, flow rate, density) refer to flow speed, i.e., harmonic mean of individual velocities, we may find the link from the individual level with the population average value in a statistical sense.

There is another equivalent approach. Let us first calculate the distribution for the average spacing of vehicles in a homogeneous platoon. Then, we could map the distribution of spacings to the distributions for density in flow-density plot. As shown in Fig. 5.5, these two approaches will get the same distributions in flow-density plot. To verify the estimated distributions, we will compare the distributions of flow rate/density that are estimated from headway distributions and the empirical flow rate/density distributions extracted from PeMS (2005).

Fig. 5.5 An illustration of the mapped distributions for points of congested flows in flow-density plot

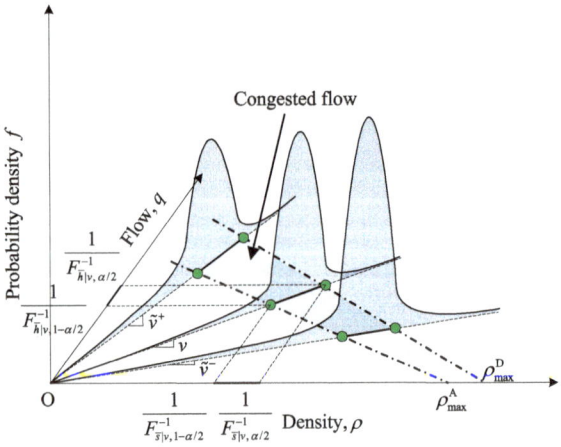

If we apply the concept of α-truncated conditional PDFs of headway proposed in Chen et al. (2010a), we could get the corresponding α-truncated conditional PDFs of flow rate. Let $F_{\bar{h}|v}$ denotes the CDF of average headways with respect to the velocity v in a vehicle platoon during a certain time interval T. We can then define $F_{\bar{h}|v,\,\alpha/2}^{-1}$ and $F_{\bar{h}|v,\,1-\alpha/2}^{-1}$ as the lower and upper confidence bounds of average headways in terms of the inverse functions of CDFs at the confidence level of α. Linking the lower and upper truncating points of each conditional PDFs of flow rates across all ranges of velocity, we can obtain the joint PDF of flow velocity that envelops a 2D region in flow-density plot. The probability that one flow-density point falls outside this 2D region is less than α. This 2D region of congested flows is more accurate than the estimation in Chen et al. (2010a).

Similarly, let $F_{\bar{s}|v,\,\alpha/2}^{-1}$ and $F_{\bar{s}|v,\,1-\alpha/2}^{-1}$ be the lower and upper confidence bounds of the average spacings in terms of the inverse functions of CDFs, we can get the same boundaries and the same 2D region of congested flows.

For the second kind of traffic flow, no definite distribution can fit average headway/spacing in a heterogeneous platoon. Thus, we cannot derive the analytical distribution of flow rate/density, correspondingly. However, we will show that the sampled points for heterogeneous platoons still locate within the α-truncated envelop that had been obtained from homogeneous platoon in flow-density plot.

5.3.2 Summation of Lognormal Random Variables

There is no closed form expression of the CDF/PDF for a sum of i.i.d. lognormal random variables. As a result, various approximations for the CDF/PDF had been proposed in the literatures (Beaulieu and Rajwani 2004; Beaulieu and Xie 2004; Ben Slimane 2001; Lam and Le-Ngoc 2006; Mehta et al. 2007; Nadarajah 2008; Zhao and Ding 2007).

Assume that n i.i.d. random variables z_i $(i = 1, 2, \ldots, n)$ follow the lognormal distribution, whose PDF is as follows:

$$f_z(z_i) = \text{Log-}\mathcal{N}(\mu_z, \, \sigma_z^2) = \frac{1}{\sqrt{2\pi}\sigma_z z_i} \exp\left[-\frac{(\log z_i - \mu_z)^2}{2\sigma_z^2}\right], \quad z_i > 0 \quad (5.6)$$

where μ_z and σ_z are the location and scale parameters, then $\log z_i$ is a normal random variable and belongs to $\mathcal{N}(\mu_z, \, \sigma_z^2)$.

Romeo et al. (2003) studied $Z_n = \sum_{i=1}^{n} z_i$, the sum of the random variables z_i and showed that

(1) if $\sigma_z^2 \ll 1$, the PDF of Z_n could be well approximated by a normal distribution.
(2) if $\sigma_z^2 \leqslant 1$, the PDF of Z_n are close to a lognormal distribution

$$f_{Z_n}(Z_n) \approx \text{Log-}\mathcal{N}(\mu_{Z_n}, \, \sigma_{Z_n}^2) \quad (5.7)$$

where μ_{Z_n} and σ_{Z_n} are the location and scale parameters of Z_n, their relations with μ_z and σ_z are as follows:

$$
\begin{cases}
\sigma_{Z_n}^2 = \log \left(1 + \frac{\exp(\sigma_z^2)-1}{n}\right) \\
\mu_{Z_n} = \log n + \mu_z + \frac{\sigma_z^2 - \sigma_{Z_n}^2}{2}
\end{cases}
\tag{5.8}
$$

The sum expectation and the typical sum can be obtained

$$
\mathrm{E}[Z_n] \approx \exp(\mu_{Z_n} + \sigma_{Z_n}^2/2), \quad Z_{n,\text{typical}} \approx \exp(\mu_{Z_n} - \sigma_{Z_n}^2) \tag{5.9}
$$

The expectation and variance of Z_n are

$$
\begin{cases}
\mathrm{E}[Z_n] = n\mathrm{E}[z_i] = n \exp \left(\mu_z + \frac{\sigma_z^2}{2}\right) \\
\mathrm{Var}[Z_n] = n\mathrm{Var}[z_i] = n \exp(2\mu_z + \sigma_z^2)(\exp(\sigma_z^2) - 1)
\end{cases}
\tag{5.10}
$$

(3) if $\sigma_z^2 \gg 1$, the PDF of the sum are very complex to find a close-form approximation.

Since we have $\sigma_z^2 \leqslant 1$ for empirical headway/spacing distributions, we will use the approximation form of lognormal distribution in our study.

5.3.3 Average Headway Distribution

In this subsection, the velocities of vehicles in a platoon are supposed to be almost the same. But we relax the strict homogeneity requirement in Newell's model and allow different spacings/headways between two consecutive vehicles.

Our previous studies based on the empirical NGSIM data and many other observations (Jin et al. 2009; Wang and Coifman 2008; Wang et al. 2009a) assumed the lognormal distributions of headway/spacing with the domain of $(0, \infty)$. However, it is not so accurate. When headway/spacing approaches to zero; the corresponding flow rate and density may exceed the traffic capacity and maximal jam density due to the reciprocals. Thus, we need to consider the infimum of spacing as the mean vehicle length and the infimum of headway as a small value larger than zero.

In order to solve this problem in reciprocal calculations, we assume that headway $h(t)$ at time t follows certain *shifted lognormal* type velocity-dependent distributions, i.e.,

$$
h(t) - h_0 \sim \text{Log-}\mathcal{N} \left(\mu_h(v(t)), \sigma_h^2(v(t))\right) \tag{5.11}
$$

where h_0 is the infimum of headway (similarly, the infimum of spacing is s_0), $\mu_h(v(t))$ is the location parameter, and $\sigma_h(v(t))$ is the scale parameter of lognormal distributions with respect to velocity $v(t)$. We will abbreviate them as μ_h and σ_h for presentation simplification.

The PDF of $h(t)$ is written as

$$f(h(t)) = \frac{1}{\sqrt{2\pi}\sigma_h(h(t) - h_0)} \exp\left[-\frac{(\log(h(t) - h_0) - \mu_h)^2}{2\sigma_h^2}\right], \quad h(t) > h_0 \tag{5.12}$$

where the expectation and the variance are

$$\begin{cases} E[h(t)] = \exp\left(\mu_h + \sigma_h^2/2\right) + h_0 \\ \text{Var}[h(t)] = \exp(2\mu_h + \sigma_h^2)(\exp(\sigma_h^2) - 1) \end{cases} \tag{5.13}$$

The so called typical value corresponds to the maximum of the PDF

$$h_{\text{typical}}(t) = \exp(\mu_h - \sigma_h^2) + h_0 \tag{5.14}$$

The coefficient that characterizes the relative dispersion of the shifted lognormal distribution is

$$C_h \equiv \sqrt{\text{Var}(h(t))}/E[h(t)] \tag{5.15}$$

Explorations on the NSGIM dataset of Highway 101 show that asymmetric driving behaviors result in a family of velocity-dependent shifted lognormal spacing/headway distributions. Among total 5,814 trajectories, we filter out the samples of motorcycles and truck automobiles, and keep the remaining 5,646 samples of cars. The average vehicle length is 4.5 m for cars. Thus, we set $s_0 = 4.5$ m, and $h_0 = 0.5$ s.[1]

As shown in Tables 5.1 and 5.2, and Fig. 5.6 μ_h decreases and μ_s increases with velocity v. The estimated parameters of μ_h, σ_h, μ_s, σ_s will be used for calculating headway/spacing distributions and corresponding flow rate/density distributions.

Suppose n vehicles in a platoon pass the loop detector sequentially. Since we assume their velocities are the same v, the sum of these n vehicles' headways measured at this point can be written as $H_n = \sum_{i=1}^{n} h_i$, where h_i are i.i.d. random variables, satisfying

$$h_i - h_0 \sim \text{Log-}\mathcal{N}\left(\mu_h, \sigma_h^2\right), \quad i = 1, 2, \ldots, n \tag{5.16}$$

where the measured headway of the ith vehicle at time t is $h_i = h_i(t)$.

---

[1] The spacing is always larger than the corresponding vehicle length. Hereafter, we select the infimum of spacing as 4.5 m. Since the infimum of headway is difficult to estimate because of varying velocities, we choose an empirical value of 0.5 s so that the headway is always larger. Actually, both the values have mild influences on the headway/spacing distributions because the estimated location and scale parameters guarantee the stable shapes of the PDFs.

Table 5.1 The estimated parameters of headway distribution from NGSIM dataset, U.S. Highway 101

v	#	$\bar{v}$	$\mu_h(v)$	$\sigma_h(v)$	$E[h(t)]$	$Var[h(t)]$	$C_h(v)$	$h_{typical}(v)$
0–3	763	2.083	1.598	0.350	5.755	3.597	0.330	4.874
3–4	1113	3.324	1.190	0.323	3.963	1.319	0.290	3.462
4–5	1924	4.552	0.970	0.338	3.293	0.943	0.295	2.854
5–6	1544	5.553	0.809	0.371	2.905	0.854	0.318	2.456
6–7	3303	6.336	0.728	0.379	2.726	0.766	0.321	2.294
7–8	5076	7.567	0.603	0.407	2.484	0.709	0.339	2.048
8–9	3682	8.538	0.499	0.440	2.314	0.703	0.362	1.857
9–10	7205	9.369	0.452	0.448	2.237	0.669	0.366	1.786
10–11	9697	10.571	0.378	0.475	2.133	0.674	0.385	1.665
11–12	4912	11.505	0.324	0.490	2.059	0.661	0.395	1.587
12–13	7734	12.360	0.290	0.501	2.015	0.655	0.402	1.539
13–14	6611	13.589	0.229	0.504	1.927	0.589	0.398	1.475
14–15	2389	14.476	0.191	0.527	1.890	0.619	0.416	1.417

Table 5.2 The estimated parameters of spacing distribution from NGSIM dataset, U.S. Highway 101

v	#	$\bar{v}$	$\mu_s(v)$	$\sigma_s(v)$	$E[s(t)]$	$Var[s(t)]$	$C_s(v)$	$s_{typical}(v)$
0–3	763	2.083	1.811	0.497	11.419	13.422	0.321	9.276
3–4	1113	3.324	2.056	0.453	13.157	17.033	0.314	10.867
4–5	1924	4.552	2.262	0.426	15.010	22.009	0.313	12.503
5–6	1545	5.553	2.359	0.437	16.135	28.437	0.330	13.241
6–7	3308	6.336	2.457	0.424	17.261	31.991	0.328	14.250
7–8	5103	7.567	2.568	0.432	18.807	41.911	0.344	15.320
8–9	3705	8.538	2.624	0.449	19.754	51.986	0.365	15.773
9–10	7296	9.369	2.702	0.441	20.938	57.980	0.364	16.780
10–11	9903	10.571	2.791	0.447	22.507	71.575	0.376	17.851
11–12	5046	11.505	2.851	0.445	23.611	79.999	0.379	18.699
12–13	8028	12.360	2.913	0.442	24.799	89.094	0.381	19.634
13–14	6977	13.589	2.978	0.431	26.070	94.991	0.374	20.824
14–15	2529	14.476	3.027	0.426	27.088	101.314	0.372	21.714

According to the analysis in Sect. 5.3.2, $H_n - nh_0$ approximately follow a log-normal style distribution. Let average headway of a platoon be defined as

$$\bar{h}_n = \frac{H_n}{n} = \frac{1}{n}\sum_{i=1}^{n} h_i \qquad (5.17)$$

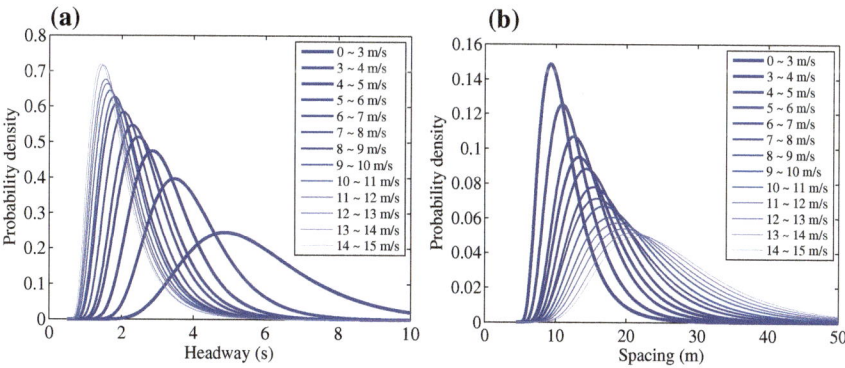

Fig. 5.6 The estimated PDFs of **a** headway and **b** spacing with respect to vehicle velocity

We have

$$\bar{h}_n - h_0 \dot{\sim} \text{Log-}\mathcal{N}\left(\mu_{H_n}(v) - \log n,\ \sigma^2_{H_n}(v)\right) = \text{Log-}\mathcal{N}\left(\mu_{\bar{h}_n}(v),\ \sigma^2_{\bar{h}_n}(v)\right) \quad (5.18)$$

whose scale and location parameters are

$$\begin{cases} \sigma^2_{\bar{h}_n}(v) = \log\left\{\frac{Var[h_i|v]}{n(E[h_i|v]-h_0)^2} + 1\right\} = \log\left(\exp(\sigma^2_h) + n - 1\right) - \log n \\ [0.3cm]\mu_{\bar{h}_n}(v) = \log\left(nE[\,h_i|\,v] - nh_0\right) - \frac{\sigma^2_{H_n}(v)}{2} - \log n = \mu_h + \frac{\sigma^2_h - \sigma^2_{\bar{h}_n}(v)}{2} \end{cases} \quad (5.19)$$

where the expectation and variance of $\bar{h}_n$ with respect to v are

$$\begin{cases} E[\bar{h}_n|\,v] = \exp(\mu_{\bar{h}_n}(v) + \sigma^2_{\bar{h}_n}(v)/2) + h_0 = \exp(\mu_h + \sigma^2_h/2) + h_0 = E[h_i|\,v] \\ [0.3cm]Var[\,\bar{h}_n|\,v] = \exp(2\mu_{\bar{h}_n}(v) + \sigma^2_{\bar{h}_n}(v))\left(\exp(\sigma^2_{\bar{h}_n}(v)) - 1\right) = Var[\,h_i|\,v]/n \end{cases} \quad (5.20)$$

According to the Hoeffding's Inequality (Hoeffding 1963) that provides an upper bound on the probability of the sum of random variables, define the upper and lower bounds of car-following headways in homogeneous platoon, i.e.,

$$\vartheta_i^- = 0 < h_i < \vartheta_i^+ = 10, \quad i \in \mathbb{N}^+ \quad (5.21)$$

where ϑ_i^- and ϑ_i^+ are the upper and lower bounds of the ith vehicle, a reasonable value of $\vartheta_i^+ = 10$ can be obtained by field data. Let ϑ_i^- and ϑ_i^+ satisfy $\vartheta^- = \sup\{\vartheta_i^-, \forall i\}$ and $\vartheta^+ = \inf\{\vartheta_i^+, \forall i\}$, then $\forall \varepsilon > 0$, we have

$$
\begin{aligned}
\Pr\left\{\bar{h}_n - \mathrm{E}\left(\bar{h}_n\right) \geqslant \varepsilon\right\} &\leqslant \Pr\left\{\max_{1\leqslant j\leqslant n}\left(H_j - \mathrm{E}\left(H_j\right)\right) \geqslant n\varepsilon\right\} \\
&\leqslant \exp\left(-\frac{2n^2\varepsilon^2}{\sum_{i=1}^{n}\left(\vartheta_i^+ - \vartheta_i^-\right)^2}\right) \\
&\leqslant \exp\left(-\frac{2n\varepsilon^2}{\left(\vartheta^+ - \vartheta^-\right)^2}\right) \leqslant \delta
\end{aligned}
\tag{5.22}
$$

Thus, the effective number of vehicles given the error limits ε and δ is

$$
n^* = -\frac{\ln\delta\left(\vartheta^+ - \vartheta^-\right)^2}{2\varepsilon^2}
\tag{5.23}
$$

Figure 5.7 shows that the differences between the true and measured values for $\bar{h}_n$ and $\bar{s}_n$ are ε_h and ε_s, respectively. The probability supremum decreases with n. This indicates that the estimation accuracy will increase with the number of vehicles. Moreover, we can also calculate the upper and lower bounds of estimate errors based on Kolmogorov's Inequality, i.e.,

$$
1 - \frac{\left(\varepsilon + 2\vartheta^+\right)^2}{\sum_{i=1}^{n}\mathrm{Var}(h_i)} \leqslant \Pr\left\{\max_{1\leqslant j\leqslant n}\left|H_j - E(H_j)\right| \geqslant \varepsilon\right\} \leqslant \frac{\sum_{i=1}^{n}\mathrm{Var}(h_i)}{\varepsilon^2}
\tag{5.24}
$$

Notice the special reciprocal property of lognormal distributions, if the mean headway distribution for each velocity range belongs to a family of lognormal type distributions, then its reciprocal shall be lognormal type as well. Thus, we can derive the PDF of the mean headway

$$
f(\bar{h}_n) = \frac{1}{\sqrt{2\pi}\,\sigma_{\bar{h}_n}(\bar{h}_n - h_0)}\exp\left[-\frac{\left(\log(\bar{h}_n - h_0) - \mu_{\bar{h}_n}\right)^2}{2\sigma_{\bar{h}_n}^2}\right], \quad \bar{h}_n > h_0
\tag{5.25}
$$

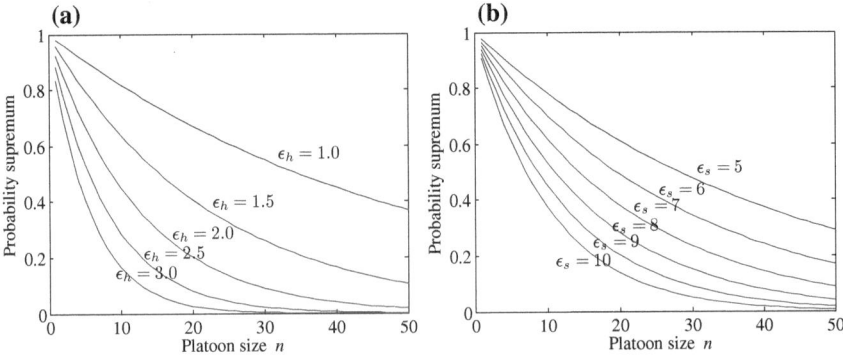

Fig. 5.7 Relationship between probability supremum and platoon size. **a** Headway $h_i \in$ (0, 10), $i \in \mathbb{N}^+$. **b** Spacing $s_i \in$ (5, 50), $i \in \mathbb{N}^+$

We assume that the platoon velocity does not change too much during a short time interval (e.g., $T = 30\,\mathrm{s}$) so that the headways belong to the same shifted lognormal distribution. The number of vehicles detected during T can then be estimated as

$$n = \left\lfloor \frac{T}{\mathrm{E}[h_i \mid v]} \right\rfloor = \left\lfloor \frac{T}{\exp(\mu_h + \sigma_h^2/2) + h_0} \right\rfloor \qquad (5.26)$$

Because the relationship is monotonic, the PDF of traffic flow rate can be derived according to $f_Q(\bar{q}) = |\mathrm{d}\bar{h}_n/\mathrm{d}\bar{q}| \, f(\bar{h}_n)$, we obtain

$$f_Q(\bar{q}) = \frac{3{,}600}{\sqrt{2\pi}\,\sigma_{\bar{h}_n}\,\bar{q}(3{,}600 - h_0\bar{q})} \exp\left[-\frac{\left(\log(3{,}600/\bar{q} - h_0) - \mu_{\bar{h}_n}\right)^2}{2\sigma_{\bar{h}_n}^2} \right]$$

$$(5.27)$$

where $f_Q(\bar{q})$ is PDF of the estimated traffic flow rate given by headway distributions.

Analogously, we could derive the distribution of density from the distribution of spacings. Based on the hypothesis that spacings also belong to a family of shifted lognormal distributions, we can derive the PDF of the mean spacing as

$$f(\bar{s}_n) = \frac{1}{\sqrt{2\pi}\,\sigma_{\bar{s}_n}(\bar{s}_n - s_0)} \exp\left[-\frac{\left(\log(\bar{s}_n - s_0) - \mu_{\bar{s}_n}\right)^2}{2\sigma_{\bar{s}_n}^2} \right], \quad \bar{s}_n > s_0 \qquad (5.28)$$

where s_0 is the infimum of spacing, in the rest of this chapter, it is chosen as the mean vehicle length.

Since the relationship between density and spacing during the predefined short time interval can be approximated as $\bar{\rho} = 1{,}000/\bar{s}_n$. The PDF of density can be derived according to $f_\rho(\bar{\rho}) = |\mathrm{d}\bar{s}_n/\mathrm{d}\bar{\rho}| \, f(\bar{s}_n)$ as the following

$$f_\rho(\bar{\rho}) = \frac{1{,}000}{\sqrt{2\pi}\,\sigma_{\bar{s}_n}\,\bar{\rho}(1{,}000 - s_0\bar{\rho})} \exp\left[-\frac{\left(\log(1{,}000 - s_0\bar{\rho}) - \log\bar{\rho} - \mu_{\bar{s}_n}\right)^2}{2\sigma_{\bar{s}_n}^2} \right]$$

$$(5.29)$$

5.3.4 Model Validation

To validate the above estimated distributions based on the vehicle trajectory dataset, we compare them with the empirical distributions of flow rate and density with respect of velocity extracted from the loop detectors (PeMS 2005; Rice and van Zwet 2004) at Vehicle Detection Station (VDS) 717490 on Highway SR101 southbound, California. Figure 5.8 shows that the loop station is at the same site where NGSIM trajectory data Highway 101 were gotten. The NGSIM trajectory data used here were collected using video cameras mounted on a 36-story building adjacent to the Lankershim Boulevard interchange. In order to evaluate the performance of

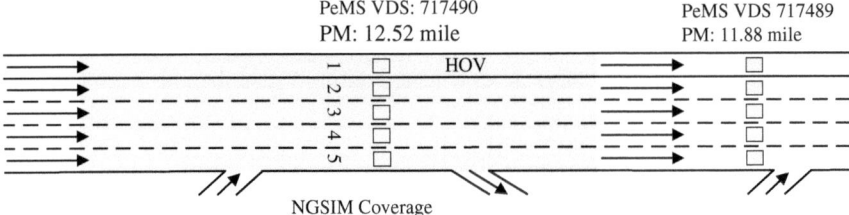

Fig. 5.8 PeMS loop detector layout and NGSIM vehicle trajectory measurement site

the proposed model aforementioned, we test more detector locations. The downstream VDS 717489 was chosen as the alternative one because the mainline layout is analogical to VDS 717490. The distance between these two locations is 0.64 mile (1.03 km), then it is reasonable to incorporate the estimated lognormal parameters of headway/spacing distributions from NGSIM dataset because the driver population doesn't vary too much at these two locations (the off-ramp may slightly reduce the population consistency). It is worthy to point out that we estimate flow rate and density distribution using different methods from two independent datasets with different data formats and other characteristics, see Table 5.3.

Thus, we can take them as different measures for the same traffic flows. Since PeMS only provides the time occupancy information, we need to choose an appropriate g-factor to retrieve the density from the time occupancy as Rice and van Zwet (2004)

Table 5.3 Characteristics of the PeMS and NGSIM datasets used in this study

Datasets	PeMS homogeneous platoon	PeMS heterogeneous platoon	NGSIM Highway 101
Locations	VDS 717490	VDS 717490	SR 101-S
		VDS 717489	
Number of lanes	5 versus middle 3	5 lanes	5 lanes
Lane-by-lane	Yes	Yes	Yes
Time intervals	30 s	5 min	0.01 s
Measurement period	06/1–30/2005	05/1–07/31/2005	7:50–8:35 am
			06/15/2005
Measurement length	30 days	92 days	45 min
Valid data length	29 days	72 days	45 min
Observed quantities	Vehicle counts	Vehicle counts	Headway
	Time occupancy	Time occupancy	Spacing
	Vehicle length	Vehicle length	Velocity
	Vehicle type	Vehicle type	Coordinates
Detected vehicles	5 lanes: 3,601,055	5 lanes: 8,196,829	6,101
	3 lanes: 2,497,869	3 lanes: 5,878,290	

Fig. 5.9 Empirical
distributions of three
vehicles' lengths

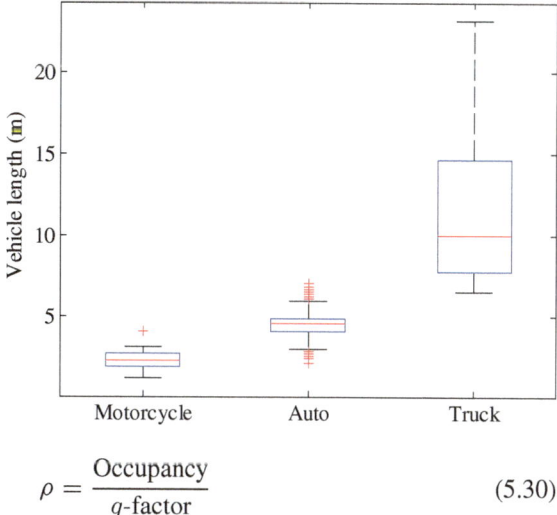

$$\rho = \frac{\text{Occupancy}}{g\text{-factor}} \qquad (5.30)$$

Figure 5.9 shows that there are three categories of vehicles in the sample. Among 5,814 sample trajectories of NGSIM Highway 101 dataset, the number of motorcycles, automobiles, and truck automobiles are 42, 5,646, and 126, respectively. And the average lengths for these vehicles are 2.39, 4.50, and 11.62 m. Noticing that almost all the vehicles are automobiles, we set g-factor as 4.50 m.

In this chapter, we categorize the empirical data points $(q, \rho, \tilde{v})$ extracted from PeMS, according to their estimated velocity $\tilde{v} = q/\rho$ instead of the time-mean velocity v fed by loop detectors. The reason on using space mean velocity is that the time-mean velocity does not coordinate well with occupancy data in the flow-density plot; see also the related discussions in Coifman et al. (2003), Dailey (1999), Li (2009, 2010), Soriguera and Robusté (2011).

As shown in Table 5.3, in order to retrieve the data for homogeneous platoons, we choose the aggregation time interval as $T = 30$ s. We assume the maximal traffic density to be $\rho_{\max} = 125$ veh/km that corresponds to the average spacing of full stop, i.e., 8 m, or the 100 % time occupancy, see the analogical measurement results in Chiabaut et al. (2010), Laval and Leclercq (2010a).

We calculate the mean time occupancies only using the data collected on the middle 3 lanes (lane 2–4) and then transfer these percentage values into traffic densities by using the relationship $\rho = \rho_{\max} \times \text{Occupancy} \times 100$ (veh/km). This is because the data collected on all five lanes include the slow traffic records of trucks and ramping vehicles, which may influence the estimated results. We will show the lane-by-lane variations by comparing the 3-lane and 5-lane results in the following sections.

In order to focus on congested flow, we only plot the points whose velocities are between 0–3, 3–4,...,14–15 m/s and densities are larger than the critical value $\rho_c = 37$ veh/km.[2] As shown in Fig. 5.10, let us divide the sector (which centers

---

[2] The critical density is the value that corresponds to the road capacity, and can be empirically estimated in the flow-density plots by using PeMS data.

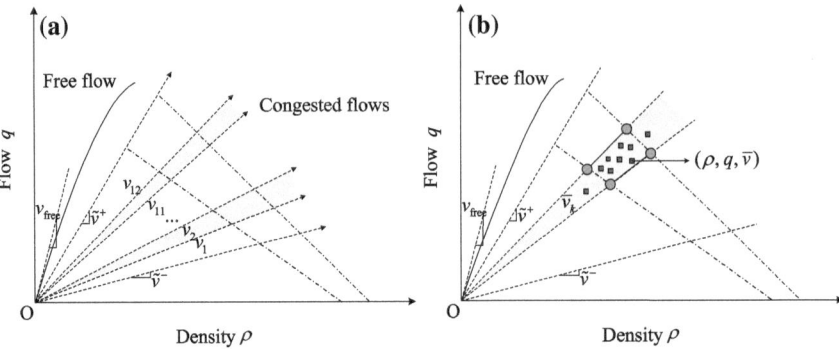

Fig. 5.10 **a** An illustration of the division plan, **b** an illustration of the division plan and the sampled flow rate and density of the points with respect to each velocity range in the flow-density plot

at the original point and the slopes of its bounds are the minimal velocity $\tilde{v}^-$ and maximum velocity $\tilde{v}^+$ in congested flow) into 13 nonoverlapping sectors according to the estimated velocity $\tilde{v} = q/\rho$. Finally, we sample q and ρ with respect to the average velocity $\bar{v}_k$ of the kth velocity range and analyze their empirical distributions.

Figure 5.11 illustrates both the estimated and empirical distributions of traffic flow rates from the aforementioned two kinds of datasets. The bars are the empirical distributions based on the sampling points extracted from the 30 s lane-aggregated PeMS data during June 1–30, 2005. The solid-line curves represent the analytical distributions calculated based on NGSIM data. The same left-skewed characteristics indicate that analytical solutions coincide with field observations in most velocity ranges. In all of these 12 velocity ranges, the results of both PDFs match each other very well. When velocity increases, the mean values of both PeMS and NGSIM datasets increase. Since we did not estimate the continuous curves by regression of field observation, the PeMS data were drawn in histograms directly and the NGSIM data were used to estimate flow rate and density PDFs, respectively. The match of loop data and trajectory data in probability distribution validate the proposed homogeneous platoon model in this book.

For the slower velocity ranges, e.g., (3, 4) and (4, 5) m/s, both the mean and variance of two datasets are highly consistent with each other. For the faster ranges, e.g., (12, 13), (13, 14), and (14, 15) m/s, the estimation of mean values are relatively more accurate. The variance of two datasets are biased mainly because $\sigma_h(v)$ increases with v in the shifted headway distributions.

Table 5.4 further gives the statistics of mean values, median values, and $\alpha = 0.05$ percentiles of flow rates for each velocity range. Though we do not have enough measurements for all the ranges, the mean flow rates well match between the simulations and observations (the differences are smaller than 6 % except for the range of 0–3 m/s).

Figure 5.12 shows the agreements between the estimated density distributions based on NGSIM data and the field measurements of PeMS, from perspectives of the mathematical expectation and the skewness direction. For most velocity ranges,

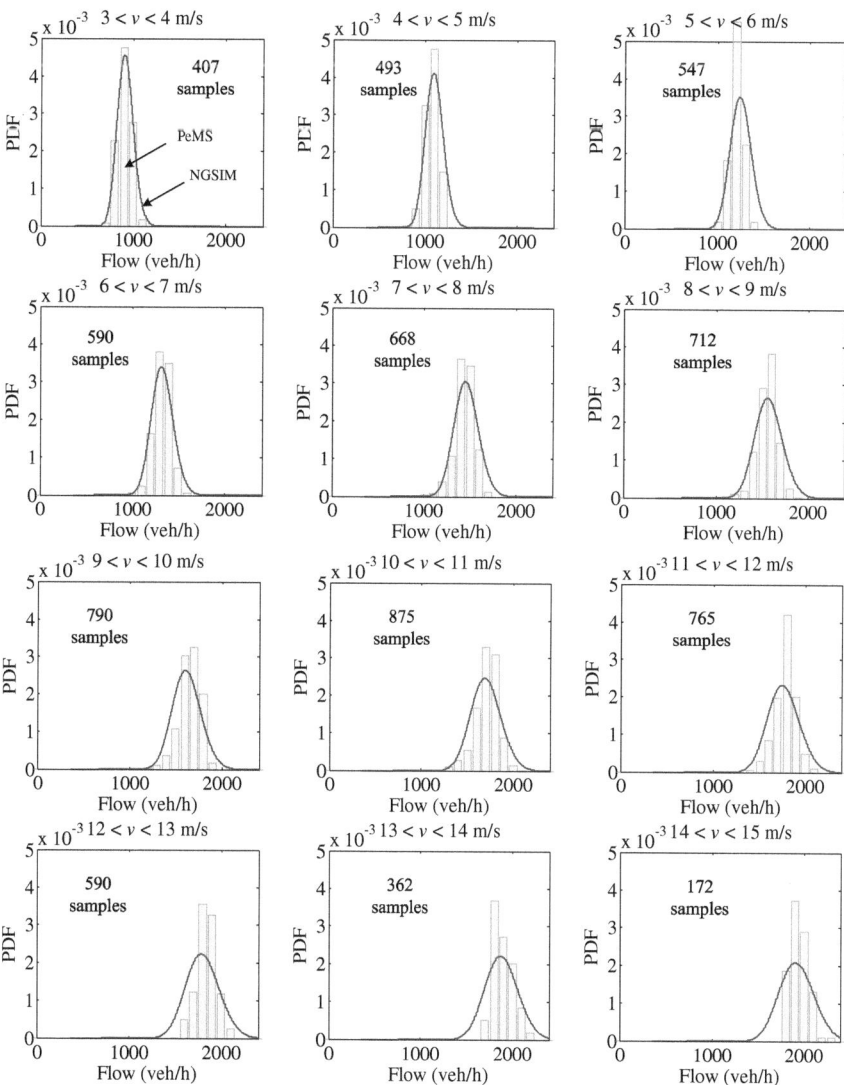

Fig. 5.11 Comparison of hypothetical and empirical distributions of traffic flow rate

the mean densities of analytical solutions match well with field observations of PeMS data. This again verifies the proposed stochastic platoon model.

For the smaller range (3, 4) m/s, the analytical density distribution from NGSIM data is right biased, because the mean spacing of small velocity range is relatively small, e.g., 8–10 m. When we calculate the analytical density distribution based on the reciprocal relationship between spacing and density, a small bias in spacing may induce a large change of density. For other smaller ranges, e.g., (4, 5) and (5, 6) m/s, the mean values are relatively accurately estimated, but the variance of two datasets

Table 5.4 Comparison of empirical (PeMS) and estimated flow rate distributions

v	#	PeMS				Analytical solution			
		Mean	Median	2.5%	97.5%	Mean	Median	2.5%	97.5%
0–3	365	760	768	528	912	635	631	494	802
3–4	536	972	984	816	1,106	917	913	754	1,100
4–5	587	1,130	1,128	960	1,292	1,102	1,098	921	1,303
5–6	676	1,277	1,272	1,090	1,440	1,250	1,246	1,038	1,485
6–7	670	1,391	1,416	1,008	1,584	1,331	1,327	1,112	1,574
7–8	1,010	1,439	1,488	1,056	1,680	1,461	1,456	1,216	1,731
8–9	1,152	1,549	1,584	1,200	1,800	1,570	1,565	1,290	1,880
9–10	1,139	1,668	1,680	1,344	1,896	1,624	1,619	1,340	1,936
10–11	870	1,758	1,776	1,398	1,992	1,703	1,698	1,400	2,036
11–12	602	1,817	1,824	1,536	2,040	1,765	1,759	1,443	2,119
12–13	330	1,874	1,872	1,608	2,112	1,804	1,798	1,470	2,172
13–14	151	1,912	1,920	1,680	2,136	1,885	1,880	1,548	2,254
14–15	53	1,971	1,968	1,820	2,160	1,924	1,917	1,565	2,317

are biased mainly because of decrease with $\sigma_s(v)$ in the shifted spacing distributions. For the faster ranges, e.g., (12, 13), (13, 14), and (14, 15) m/s, both the mean and variance of two datasets are highly consistent with each other.

Table 5.5 shows the statistics of mean values, median values, and $\alpha = 0.05$ percentiles of densities for each velocity range. All the differences in mean values and median values of densities are smaller than 4 and 15%, respectively. This again verifies the proposed stochastic platoon model.

To quantitate the statistical difference of the estimated continuous PDFs based on NGSIM data and the directly measured discrete histograms based on PeMS data shown in Figs. 5.11 and 5.12, we calculate the expectation and standard deviation of the continuous random variable $\bar{q}$ (flow rate) as follows:

$$\mathrm{E}_{\mathrm{NGSIM}}[\bar{q}] = \int_0^\infty f_Q(\bar{q})\bar{q}\,\mathrm{d}\bar{q} \tag{5.31}$$

$$\mathrm{SD}_{\mathrm{NGSIM}}[\bar{q}] = \sqrt{\int_0^\infty f_Q(\bar{q})\bar{q}^2\,\mathrm{d}\bar{q} - \left(\int_0^\infty f_Q(\bar{q})\bar{q}\,\mathrm{d}\bar{q}\right)^2} \tag{5.32}$$

where $\mathrm{E}_{\mathrm{NGSIM}}$ is the expectation of the flow rate based on NGSIM data, $\mathrm{SD}_{\mathrm{NGSIM}}$ is the standard deviation of the flow rate, $f_Q(\bar{q})$ is the flow rate PDF, similar expressions of density can be obtained, and are omitted in this book.

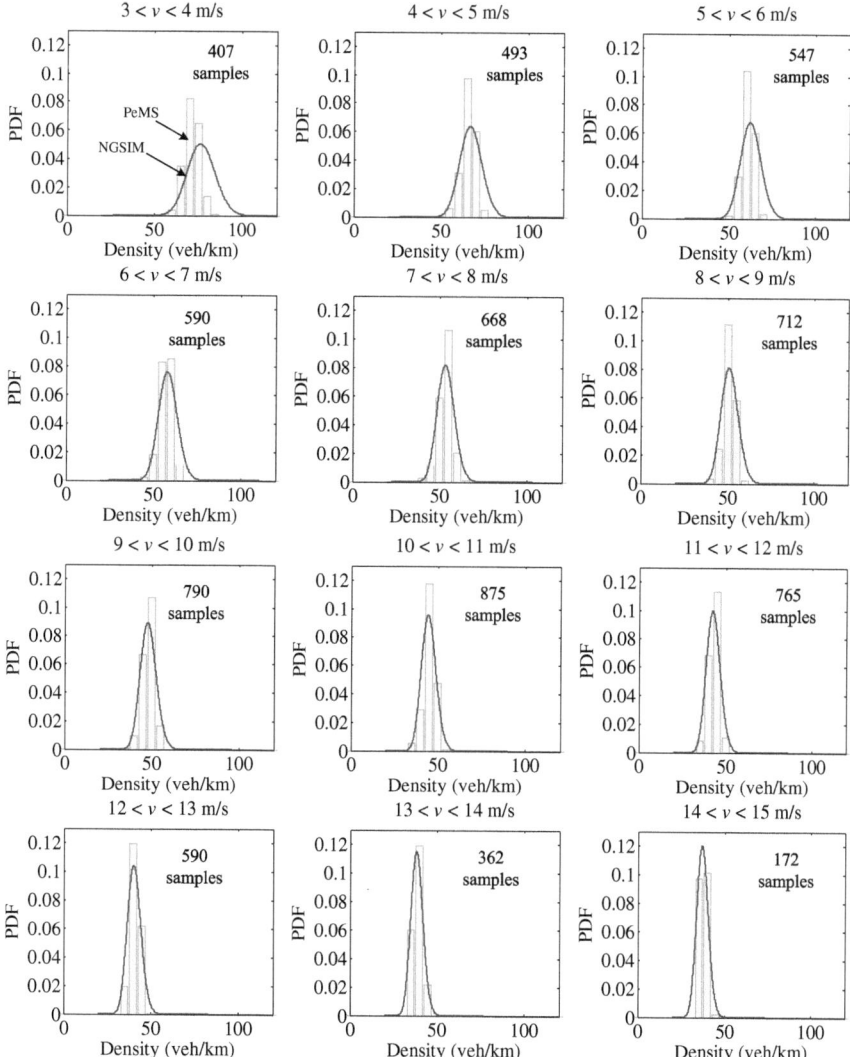

Fig. 5.12 Comparison of hypothetical and empirical distributions of density

$$\mathrm{E}_{\mathrm{PeMS}}[\bar{q}] = \frac{1}{n_k} \sum_{i=1}^{n_k} q_{i,k} \tag{5.33}$$

$$\mathrm{SD}_{\mathrm{PeMS}}[\bar{q}] = \sqrt{\frac{1}{n_k} \sum_{i=1}^{n_k} (q_{i,k} - \mathrm{E}_{\mathrm{PeMS}}[\bar{q}])^2} \tag{5.34}$$

Table 5.5 Comparison of empirical (PeMS) and estimated traffic density distributions

v	#	PeMS				Analytical solution			
		Mean	Median	2.5 %	97.5 %	Mean	Median	2.5 %	97.5 %
0–3	365	87	86	76	102	89	89	69	110
3–4	536	77	77	67	86	77	77	62	93
4–5	587	70	70	61	78	67	67	55	80
5–6	676	64	65	55	72	63	62	51	75
6–7	670	59	60	42	67	58	58	48	69
7–8	1,010	53	55	40	62	54	53	44	64
8–9	1,152	51	52	40	59	51	51	42	61
9–10	1,139	49	49	39	56	48	48	40	57
10–11	870	47	47	37	53	45	45	37	53
11–12	602	44	45	37	50	43	43	35	51
12–13	330	42	42	36	47	41	41	33	49
13–14	151	40	40	35	44	39	39	32	46
14–15	53	38	38	35	42	37	37	31	44

where E_{PeMS} is the mean flow rate from directly measured 30 s vehicle counts of PeMS data, n_k is the number of points in the kth velocity range, $q_{i,k}$ is the flow rate of the ith point in the kth velocity range, SD_{PeMS} is the standard deviation of the flow rate of PeMS data, similar expressions of density can be obtained and are omitted in this book.

Figure 5.13a shows the comparison of mean flow rates between PeMS data and NGSIM data, and their changing trends with velocity. The mean flow rates monotonously increase with velocity. The error bars show positive and negative of stan-

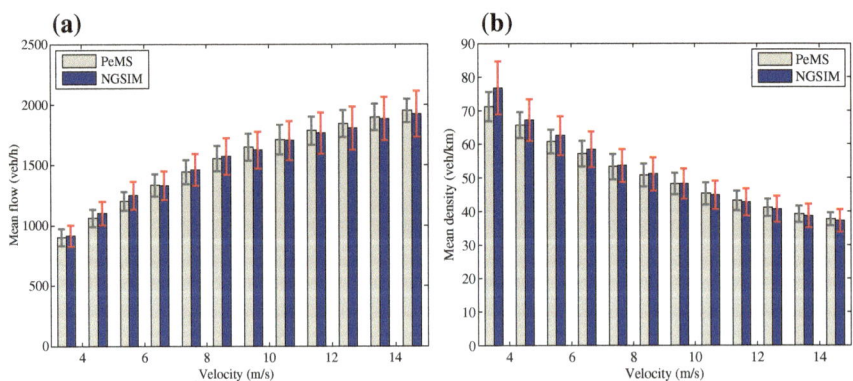

Fig. 5.13 Comparison of mean and standard deviation of hypothetical and empirical distributions. **a** Flow rate. **b** Density

dard deviations. The average maximal absolute relative biases of the mean flow rate and density for each velocity range are 1.55 and 1.85 %, respectively. The maximal absolute relative bias of the mean flow rate is 3.82 % (the 3rd velocity range, 5–6 m/s, i.e., $|1250 - 1204|/1204 = 3.82$ %). Figure 5.13b shows the comparison of mean densities between PeMS data and NGSIM data. The mean densities monotonously decrease with velocity. The maximal absolute relative bias of the mean density is 7.75 % (the 1st velocity range, 3–4 m/s, i.e., $|76.8 - 71.3|/71.3 = 7.75$ %). Figure 5.13 also shows that the NGSIM data based standard deviations are a little large than the PeMS data based values, but the standard deviations are quite small compared to the mean values. The consistency of mean values of flow rate and density validate the proposed homogeneous platoon model based on PeMS and NGSIM data.

5.3.5 Sensitivity Analysis

Several factors may influence the estimation and comparison results in this chapter, including lane-by-lane variations (see Chen et al. 2010c), velocity intervals (number of intervals and velocity ranges), and the so called g-factor. We will conduct a sensitivity analysis for these factors in this section to reveal their effects.

Figure 5.14 shows the significant differences of flow rate versus density relationship for five lanes at two detection stations on SR 101-S. Lanes 1 and 5 are the HOV lane and weaving lane, respectively, their flow-density relationships do not show obvious transitions between free states to less congested states (capacity is not achieved). Lanes 2–4 are relatively stationary lanes with obvious capacity drop. This clearly reflects the lane-by-lane variability of the density-flow relationship. All five lanes include diverse traffic of trucks and slow traffic for exiting and entering. This may influence on the estimated result. It might be better to use some lanes

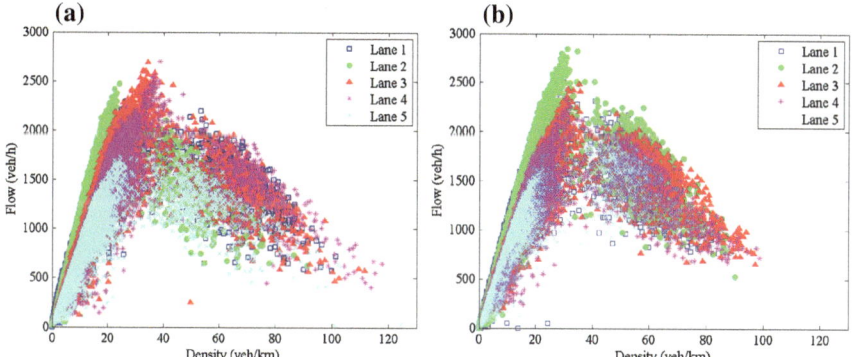

Fig. 5.14 Lane-by-lane variability of density-flow relationships across all lanes SR101-S Freeway, California, Jun 1–30, 2005. **a** VDS 717490. **b** VDS 717489

(for example, lane 1–3) only to have reliable result. We will compare the estimation results using 5 lanes and middle 3 lanes in the followings.

To quantitate the difference between the CDFs based on PeMS and NGSIM datasets, we incorporate the nonparametric K-S statistic to compare a sample with a reference probability distribution. It measures the maximal vertical separation between the empirical distribution function of the sample and the CDF of the reference distribution. We define the K-S statistic for a given flow rate CDF

$$\text{KS}_{Q,k} = \sup_{\bar{q}} |F_{Q,k,\text{ PeMS}}(\bar{q}) - F_{Q,k,\text{ NGSIM}}(\bar{q})| \tag{5.35}$$

where $\text{KS}_{Q,k}$ is the K-S statistic for flow rate distribution of the kth velocity range, $F_{Q,k,\text{ PeMS}}(\bar{q})$ is the empirical distribution function of flow rate based on n_k points in the kth velocity range based on PeMS datasets, $F_{Q,k,\text{NGSIM}}(\bar{q}) = \int_0^{\bar{q}} f_Q(\bar{q})\mathrm{d}\bar{q}$ is the estimated CDF of flow rate by NGSIM datasets for the kth velocity range, sup is the supremum of the set of distances.

Similarly, we have the K-S statistic for a given density CDF

$$\text{KS}_{\rho,k} = \sup_{\bar{\rho}} |F_{\rho,k,\text{ PeMS}}(\bar{\rho}) - F_{\rho,k,\text{ NGSIM}}(\bar{\rho})| \tag{5.36}$$

where $\text{KS}_{\rho,k}$ is the K-S statistic, $F_{\rho,k,\text{ PeMS}}(\bar{\rho})$ is the empirical CDF, $F_{\rho,k,\text{NGSIM}}(\rho) = \int_0^{\bar{\rho}} f_\rho(\bar{\rho})\mathrm{d}\bar{\rho}$ is the given CDF.

Take 12 velocity ranges as an illustration of how different lane grouping strategies influence the results in accuracy. Table 5.6 compares the K-S statistics of using middle 3 lanes and summing up all 5 lanes. The former one mainly considers the stationary

Table 5.6 Comparison of K-S statistics of two lane grouping strategies, 12 velocity ranges

k	v	Middle lanes (lane 2–4)			All lanes (lane 1–5)		
		Sample	$\text{KS}_{Q,k}$	$\text{KS}_{\rho,k}$	Sample	$\text{KS}_{Q,k}$	$\text{KS}_{\rho,k}$
1	3–4	407	0.137	**0.363**	404	**0.123**	0.403
2	4–5	493	**0.229**	**0.165**	512	0.251	0.193
3	5–6	547	**0.223**	**0.168**	574	0.249	0.228
4	6–7	590	0.173	**0.127**	633	**0.141**	0.172
5	7–8	668	**0.111***	**0.084***	646	0.140*	0.109*
6	8–9	712	**0.120**	**0.089**	772	0.162	0.121
7	9–10	790	**0.150**	**0.063**	818	0.192	0.119
8	10–11	875	0.160	**0.135**	892	**0.142**	0.149
9	11–12	765	0.185	0.152	770	**0.176**	**0.146**
10	12–13	590	0.270	0.183	545	**0.230**	**0.154**
11	13–14	362	0.211	0.223	301	**0.185**	**0.167**
12	14–15	172	0.339	0.277	122	**0.276**	**0.266**

* means the minimum value of each column

lanes where traffic flow are least impacted by lane management and on/off-ramps, while the latter one ignore the lane-by-lane variations. The bold numbers show the smaller value of K-S statistic between two strategies. The stars mean the minimal KS_Q or KS_ρ among all velocity ranges for each strategy. We can find that for smaller velocity ranges, i.e., (3–10) m/s, using middle 3-lane aggregated data generate better estimation accuracy for both flow rate and density, while for larger velocity ranges, i.e., (11–15) m/s, using all 5-lane aggregated data generate better estimation accuracy for both flow rate and density.

In order to investigate the effects of different velocity ranges on the results, we define the following weighted mean K-S statistics with the consideration of sample size of each velocity range

$$\overline{KS}_Q = \frac{\sum\limits_k n_k KS_{Q,k}}{\sum\limits_k n_k}, \quad \overline{KS}_\rho = \frac{\sum\limits_k n_k KS_{\rho,k}}{\sum\limits_k n_k} \tag{5.37}$$

Figure 5.15 shows the comparisons between PeMS data based empirical CDF and NGSIM data based analytical CDF of flow rates and density when 5 equal velocity ranges are chosen in (3, 15) m/s, each range size of which is 2.4 m/s.

We compare 5, 10, 12, 15, 20 different but equally disaggregated velocity ranges in this chapter to check if a different disaggregation period would change the estimated distributions. Table 5.7 shows the influence of the velocity range threshold. The starred number represents the minimal value of each column. Results show that the 5 velocity ranges perform better than other disaggregation plans. Once again, most K-S statistics of middle 3 lanes (Lane 2–4) aggregation is smaller than all 5 lanes (Lane 1–5) aggregation. We could also find that $\overline{KS}_\rho < \overline{KS}_Q$ holds for all disaggregation plans, because the hourly flow rate is estimated by multiplying 30 s vehicle counts by 120, while the density is calculated based on 30 s time occupancies. The former one may result in a large change in hourly flow rate, e.g., 120 veh/h, if

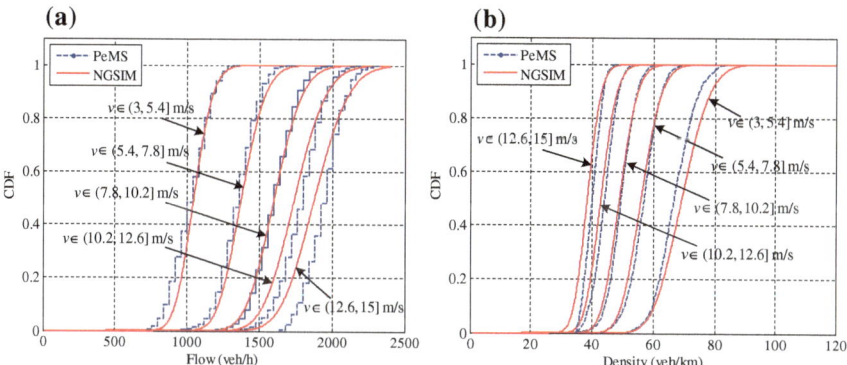

Fig. 5.15 Comparison of CDFs based on PeMS and NGSIM datasets. **a** Flow rate. **b** Density

Table 5.7 The weighted average K-S statistics for different velocity intervals

Velocity intervals	Middle lanes (lane 2–4)		All lanes (lane 1–5)	
	$\overline{KS}_Q$	$\overline{KS}_\rho$	$\overline{KS}_Q$	$\overline{KS}_\rho$
5	0.153*	**0.124***	**0.131***	0.129*
10	0.171	**0.147**	**0.169**	0.165
12	**0.179**	**0.149**	0.180	0.168
15	**0.186**	**0.143**	0.193	0.165
20	**0.195**	**0.154**	0.199	0.177

* means the minimum value of each column

one vehicle changes in the 30 s counts, but the latter one is not so sensitive with field measurements.

To illustrate the influence of g-factor that link the time occupancy with density (Occupancy $\times$ 100 % $\times$ 1,000/g-factor = Density) on estimation results. The g-factor (effective vehicle length plus effective detector length) is used to calculate the average vehicle speeds from the flow and occupancy data. Typically, PeMS uses an adaptive algorithm to compute the g-factor per each loop to provide accurate speed estimates because g-factor varies by lane, time of day, as well as the loop sensitivity. However, it is usually difficult to obtain the time-varying g-factors in practice. We try to test several typical values and their sensitivity to the results. Tables 5.8 and 5.9 list different values of g-factors, i.e., 5.1, 5.2, 5.3, 5.4, 5.5 m, we could find the best g-factor, which induces the minimal weighed K-S statistics for both flow rate and density, is 5.3 m. To roughly find the regulation from the results, for slower ranges, a smaller g-factor may increase the accuracy of estimation results (reduced CDF difference), while for faster ranges, a larger g-factor may increase the accuracy of estimation results.

Table 5.8 Effects of different g-factors, $KS_{Q,k}$ for flow rate distributions

k	v	g-factor 5.5 m	g-factor 5.4 m	g-factor 5.3 m	g-factor 5.2 m	g-factor 5.1 m
1	3–4	0.236	0.194	**0.137**	0.144	0.175
2	4–5	0.346	0.287	0.229	0.182	**0.127**
3	5–6	0.340	0.291	0.223	0.165	**0.114***
4	6–7	0.171	**0.121***	0.173	0.222	0.259
5	7–8	0.183	0.139	**0.111***	0.160	0.216
6	8–9	0.201	0.164	**0.120**	0.118*	0.172
7	9–10	**0.108***	0.123	0.150	0.179	0.234
8	10–11	**0.117**	0.137	0.160	0.194	0.228
9	11–12	**0.124**	0.168	0.185	0.206	0.241
10	12–13	**0.207**	0.230	0.270	0.283	0.297
11	13–14	0.212	0.215	**0.211**	0.240	0.248
12	14–15	**0.315**	0.337	0.339	0.387	0.318
$\overline{KS}_Q$		0.194	0.183	**0.179**	0.194	0.216

* means the minimum value of each column

Table 5.9 Effects of different g-factors, $KS_{\rho,k}$ for density distributions

k	v	g-factor 5.5 m	g-factor 5.4 m	g-factor 5.3 m	g-factor 5.2 m	g-factor 5.1 m
1	3–4	0.445	0.406	0.363	0.309	**0.257**
2	4–5	0.267	0.222	0.165	0.126	**0.096**
3	5–6	0.256	0.221	0.168	0.123	**0.109**
4	6–7	0.223	0.177	0.127	0.089*	**0.077***
5	7–8	0.139	0.091	**0.084**	0.142	0.174
6	8–9	0.135	0.091	**0.089**	0.113	0.142
7	9–10	0.113	0.079*	**0.063***	0.101	0.120
8	10–11	**0.084***	0.103	0.135	0.185	0.218
9	11–12	**0.087**	0.128	0.152	0.169	0.202
10	12–13	**0.108**	0.135	0.183	0.194	0.241
11	13–14	**0.209**	0.219	0.223	0.227	0.256
12	14–15	**0.276**	0.278	0.277	0.316	0.267
$\overline{KS}_\rho$		0.171	0.155	**0.149**	0.159	0.171

* means the minimum value of each column

5.4 The Heterogeneous Platoon Model

5.4.1 Average Headway Distribution

In practice, we often encounter complex traffic dynamics which cannot be simply modeled as a homogeneous platoon. In the following discussion, we will show that average headway/spacing in a heterogeneous platoon does not follow a definite distribution. However, we will also prove that the points for heterogeneous platoons still locate within the α-truncated envelop for homogeneous platoons.

Proposition 5.2 *Given a heterogeneous platoon $\tilde{\Upsilon}$ consisting of n vehicles, it can be absolutely decomposed into m homogeneous subplatoons Υ_k, $k = 1, 2, \ldots, m$, satisfying $\bigcup_{k=1}^{m} \Upsilon_k = \tilde{\Upsilon}$ and $\bigcap_{k=1}^{m} \Upsilon_k = \emptyset$, $\forall k$, if Υ_k keeps in congested flows, then $\tilde{\Upsilon}$ is located within the probabilistic boundaries of congested flows as well.*

Proof Mathematical induction is incorporated to prove this proposition as follows.

When $m = 1$, the previously defined probabilistic boundaries of congested flows are established based on the assumption of homogeneous and stable platoons. In this simple case, $\tilde{\Upsilon}$ is a homogeneous platoon, indicating that the speeds of n vehicles are in the same range, so the above proposition clearly holds.

When $m = 2$, without loss of generality, assume $\tilde{\Upsilon}$ can be absolutely decomposed into two homogeneous subplatoons Υ_1 and Υ_2, consisting of n_1 and n_2 vehicles, satisfying $n_1 + n_2 = n$. When Υ_1 and Υ_2 pass through the detecting section with mean velocities of v_1 and v_2, respectively, the passing times are T_1 and T_2, let

$T_1 + T_2 = T$. Since the different speeds of the two subplatoons, $\bigcup_{k=1}^{m} \Upsilon_k = \tilde{\Upsilon}$ is inhomogeneous.

Suppose the mean headways of Υ_1 and Υ_2 are $\bar{h}_1$ and $\bar{h}_2$, respectively. Thus, the passing number of vehicles in the first platoon Υ_1 is $n_1 = T_1/\bar{h}_1$, and the passing number of vehicles in the second platoon Υ_2 is $n_2 = T_2/\bar{h}_2$. The mean velocity and mean headway of this larger platoon measured at this point is

$$\tilde{v} = \frac{v_1 n_1 + v_2 n_2}{n_1 + n_2} = \frac{v_1 T_1 \bar{h}_2 + v_2 T_2 \bar{h}_1}{T_1 \bar{h}_2 + T_2 \bar{h}_1} \tag{5.38}$$

$$\tilde{h} = \frac{T_1 + T_2}{n_1 + n_2} = \frac{\bar{h}_1 \bar{h}_2 (T_1 + T_2)}{T_1 \bar{h}_2 + T_2 \bar{h}_1} \tag{5.39}$$

Suppose the lower and upper boundary points of headway distributions are h_1^- and h_1^+ for velocity v_1, and h_2^+ and h_2^- for velocity v_2, satisfying $h_1^- \leqslant \bar{h}_1 \leqslant h_1^+$ and $h_1^- \leqslant \bar{h}_2 \leqslant h_2^+$. As shown in Fig. 5.16, all the lower (upper) boundary points fall in one line, then we have the lower and upper boundary points of headway distributions for velocity $\tilde{v}$ as

$$\tilde{h}^- = \frac{h_1^- h_2^- (T_1 + T_2)}{T_1 h_2^- + T_2 h_1^-}, \quad \tilde{h}^+ = \frac{h_1^+ h_2^+ (T_1 + T_2)}{T_1 h_2^+ + T_2 h_1^+} \tag{5.40}$$

Without loss of generality, we assume $v_1 < v_2$, then $h_1^- > h_2^-$ and $h_1^+ > h_2^+$. Thus, we have

$$h_2^- < \tilde{h}^- < h_1^-, \quad h_2^+ < \tilde{h}^+ < h_1^+ \tag{5.41}$$

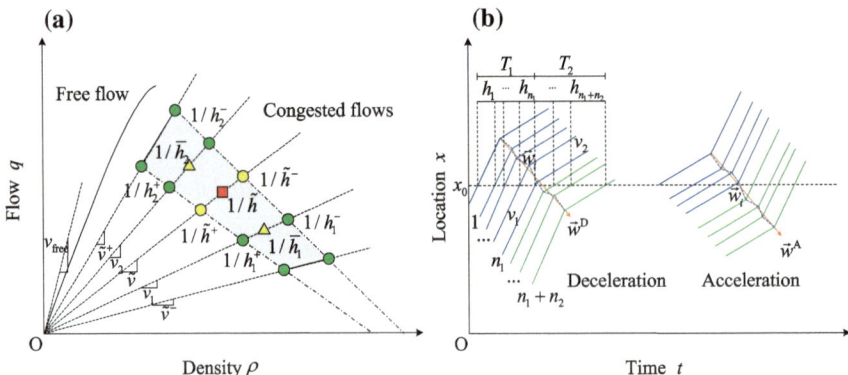

Fig. 5.16 a An illustration of the average headway of a heterogeneous platoon in flow-density plot; **b** an illustration of the trajectories of this mixed platoon, where x denotes the location of the loop detector. We can see that the speed of the platoon changes, because an upstream shock front (congestion wave) passes this loop detector during measuring

Furthermore,

$$\tilde{h}^- \leqslant \tilde{h} \leqslant \tilde{h}^+ \tag{5.42}$$

So, the points denoting this heterogeneous platoon $\tilde{\Upsilon}$ are still within the α-truncated envelop that had been determined from homogeneous platoons in the flow-density plot. Using the same decomposition trick, we can extend this conclusion to general platoons, $\forall m > 2$ containing even more complex traffic dynamic.

In summary, the confidence level is not only a parameter of stochastic distribution but also a physical parameter of traffic flow. On the microscopic level, it depicts the preference of drivers on spacings and headways; on the macroscopic level, it characterizes the boundaries that are obtained via distribution models and thus provides the probability that an observed point falls outside the congested flow region.

5.4.2 Validation

To check whether the predicated boundaries envelop the points in flow-density plot, we still resort to empirical PeMS data; but the aggregation time interval is set as $T = 5$ min. For each range of v (expect the low-velocity range between 0 and 3 m/s because of the relatively small sample size), we estimate the probabilistic boundaries based on the homogeneous platoon model as follows:

Define the CDF of $h(t)$ as $F_{h|v}$ with respective to v. The $\alpha/2$ lower bounds and the $(1 - \alpha/2)$ upper bounds of $h(t)$ can then be denoted as

$$F_{h|v}(\exp(\mu_h - \sigma_h \eta_{\text{lower}}) + h_0) = \frac{\alpha}{2} \quad F_{h|v}(\exp(\mu_h - \sigma_h \eta_{\text{upper}}) + h_0) = 1 - \frac{\alpha}{2} \tag{5.43}$$

the inverse function are

$$F_{h|v,\,\alpha/2}^{-1} = \exp(\mu_h - \sigma_h \eta_{\text{lower}}) + h_0, \quad F_{h|v,\,1-\alpha/2}^{-1} = \exp(\mu_h - \sigma_h \eta_{\text{upper}}) + h_0 \tag{5.44}$$

where η_{lower} is the $\alpha/2$ quantile and η_{upper} is $(1 - \alpha/2)$ quantile of the standard normal distribution $\mathcal{N}(0, 1)$, i.e., $\eta_{\text{lower}} = \Phi^{-1}(\alpha/2)$ and $\eta_{\text{upper}} = \Phi^{-1}(1 - \alpha/2)$, and $\Phi^{-1}(\cdot)$ is the inverse CDF of $\mathcal{N}(0, 1)$.

Thus, the mapped boundary points for $\bar{h}_n$ are

$$q_{\text{upper}}(v, n) = F_{Q|v,\,\alpha/2}^{-1} = \frac{1}{F_{\bar{h}_n|v,\,\alpha/2}^{-1}} \tag{5.45}$$

$$q_{\text{lower}}(v, n) = F_{Q|v,\,1-\alpha/2}^{-1} = \frac{1}{F_{\bar{h}_n|v,\,1-\alpha/2}^{-1}} \tag{5.46}$$

where $q_{lower}(v)$ and $q_{upper}(v)$ are the minimum/maximum flow rate, analogously, $\rho_{lower}(v)$ and $\rho_{upper}(v)$ are the minimum/maximum density, $F_{Q|v} = \int f_Q(\bar{q})d\bar{q}$ is the CDF of average flow rate.

Since we get $q_{lower}(v) = v\rho_{lower}(v)$ and $q_{upper}(v) = v\rho_{upper}(v)$, the boundary of the 2D region of congested flow can finally be calculated as

$$\bigcup_{\forall v} (\rho_{lower}(v), \ q_{lower}(v)), \quad \bigcup_{\forall v} (\rho_{upper}(v), \ q_{upper}(v)) \tag{5.47}$$

with $v \in [\tilde{v}^-, \tilde{v}^+]$.

We still apply the same values of n as what had been used in Figs. 5.11 and 5.12. Linking these upper and lower confidence bounds by linear regression method, respectively, we obtain an envelope of congested flows for each particular confidence level α. We have the following hypothesis test problem:

$$\begin{aligned} H_0 &: (\rho_i, \ q_i) \in \text{Envelop} \\ H_1 &: (\rho_i, \ q_i) \notin \text{Envelop} \end{aligned} \tag{5.48}$$

where $(\rho_i, \ q_i)$ is the estimated density-flow rate points by PeMS datasets in the flow-density plane.

Figure 5.17 shows the empirical flow-density plot from the same loop detector (PeMS 2005), where the vehicle count and time occupancy data (aggregated from 5 lanes in the mainline) were collected from May 1 to July 31, 2005 (three months). We can see that for a small enough value of α, e.g., $\alpha = 0.01$, most scattering points for the congested flows locate in the estimated envelope region.

The estimated envelop region shrinks when α increases, and the percentage of points outside the envelop decreases simultaneously. Given a truncating threshold α,

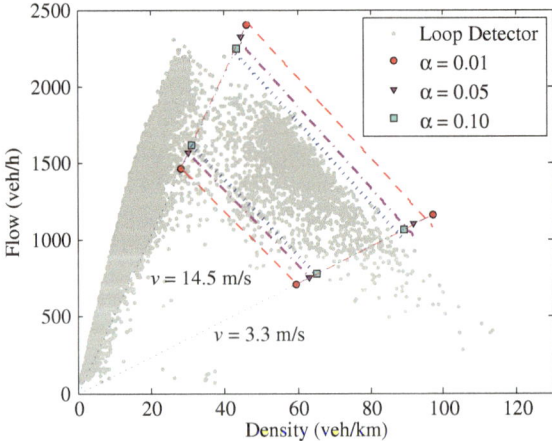

Fig. 5.17 The estimated probabilistic boundaries of congested flow with respect to the α values

we may predict that at least $(1 - \alpha) \times 100\%$ points will fall into the envelope region. Table 5.5 compares the empirical percentage of points that fall into the envelope region and the predicted percentage under difference confidence levels. The accuracy of the estimated probabilistic boundaries are validated by comparing the statistical results of scatter PeMS data with the envelope region. For all selected α (we take 0.01, 0.05, 0.10, 0.15, and 0.20 as examples here), the percentages of scattering points lie outside the boundaries are all smaller than $\alpha \times 100\%$. This agreement supports the proposed stochastic platoon model, again.

5.4.3 Boundaries of Congested Flows

Field data in Tables 5.1 and 5.2 show that $\mu_h(v)$ decreases while $\mu_s(v)$ increases with vehicle velocity v, until reaches free velocity v_{free}. Roughly, the probabilistic boundary points of congested flows are approximately on two straight lines with negative slopes.

Noticing another famous assumption that the congested points lie in one sector (centers at the original point and the slopes of its bounds are the minimal flow velocity $\tilde{v}^-$ and maximum flow velocity $\tilde{v}^+$) (Forbes 1963), we finally get a trapezoid boundary curve for the 2D region of congested flows.

Figure 5.18 and Table 5.10 show the estimated trapezoid 2D region with respect to different significance level α and vehicle numbers n. According to the concentration property for the sum of i.i.d. random variables (Petrov 1975) the variance of $\bar{h}_n$, H_n will decrease with n. Thus, as shown in Fig. 5.18, the 2D region of congested flows will shrink with the increase of n.

It is worthy to notice that the probabilistic boundary lines do not cross the horizontal axis at the same point. Differently, these two lines were assumed to intersect the same point $(\rho_{\text{max}}, 0)$ in Chen et al. (2010b), Forbes (1963), Kim and Zhang (2008). One possible answer to this difference is, from the viewpoint of spacing distribution, minimum (jam) spacing is also a random variable, if the heterogeneity of drivers is considered. Some discussions on the spacing distributions for parked vehicles could be found in Jin et al. (2009) and the references within. As a result, ρ_{max}^A and ρ_{max}^D are different. ρ_{max}^D corresponds with the jam spacing chosen by the boldest driver and ρ_{max}^A corresponds with the jam spacing chosen by the most cautious driver.

It is hard to judge which assumption is better at present, since we still do not have enough data for heavily congested traffic flow. In almost all studies, e.g., Li and Zhang (2011), we can just get one or few sparse points located in the right bottom corner of the FD. The lack of data prevents us from verifying the feature of flow-density plot, when the density approaches the maximum value. We hope that new data collected in severe congestions could help solving this puzzle in the future.

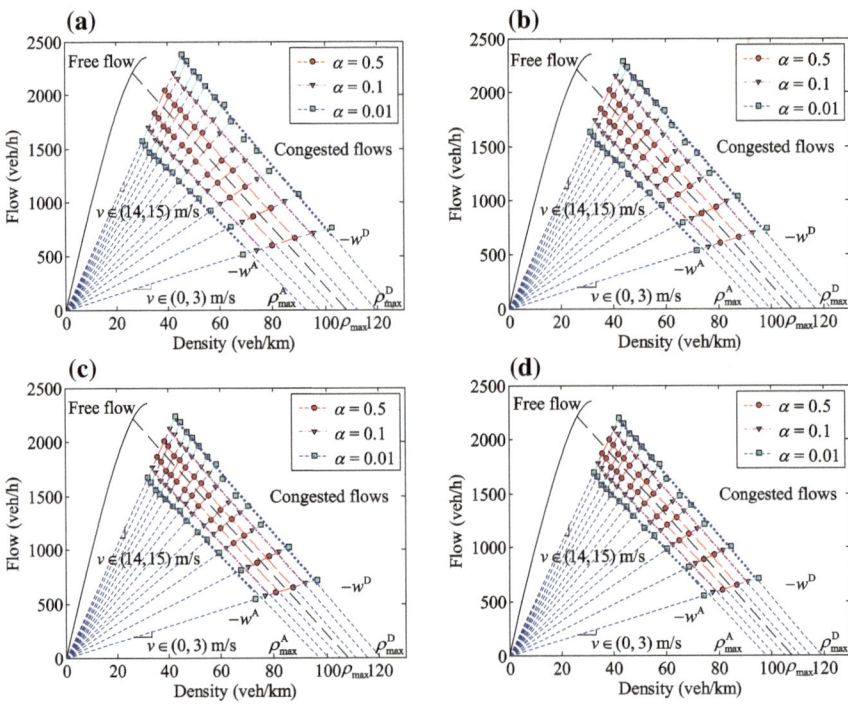

Fig. 5.18 Comparison of probabilistic boundaries of congested flows in the flow-density plot at different significance levels based on headway **a** $n = 20$ **b** $n = 30$ **c** $n = 40$ **d** $n = 50$

Table 5.10 Statistical validation of probabilistic boundaries using PeMS datasets

Location	Confidence level α	Inside envelope (%)	Outside envelope	Accepted hypothesis
Upstream	0.01	99.38	0.62% < 0.01	H_0
VDS 717490	0.05	97.77	2.23% < 0.05	H_0
SR101-S, CA	0.10	96.01	3.99% < 0.10	H_0
	0.15	93.92	6.08% < 0.15	H_0
	0.20	90.46	9.54% < 0.20	H_0
Downstream	0.01	99.28	0.72% < 0.01	H_0
VDS 717489	0.05	97.54	2.46% < 0.05	H_0
SR101-S, CA	0.10	93.23	6.77% < 0.10	H_0
	0.15	85.77	14.23% < 0.15	H_0
	0.20	78.32	21.68% < 0.20	H_1

5.5 Summary

Due to unavoidable randomness of driving actions, we observe the sampling points of loop data spread in the flow-density plot. Some researchers had attempted to interpret the scattering features based on spacing/headway variations. However, to the best of our knowledge, no clear mathematical models had been proposed to fill this void in previous studies.

In this chapter, we discuss the relationships between the spacing/headway distributions, the microscopic spacing-velocity plot, and also the macroscopic flow-density plot. Particularly, we propose three new viewpoints in our study:

First, we extend the conventionally deterministic Newell's simplified car-following model into stochastic models and allow random headways/spacings in a homogeneous platoon in which vehicles run at the same velocity.

Second, we extent the conventionally deterministic reciprocal relation between headway and flow rate into a stochastic form: *the reciprocal of average headway of vehicles in a homogeneous platoon and the corresponding flow rate should follow the same distribution.* From this way, we could connect average headways in a platoon that pass the loop detector and the flow rate measured in the same period. In other words, we can link microscopic headways/spacings and macroscopic flow rates/densities through this stochastic platoon model.

Third, we interpret the flow-density plot by emphasizing the hidden variable "velocity". Under the orthogonal coordinates of density and flow rate, the points seem disorderly scattering. But when we revisit these points under either the polar coordinates of velocity and flow rate or the polar coordinates of velocity and density, the hidden distribution tendency emerges.

We find that when the aggregation time interval is as short as 30 s, the measured vehicles could be viewed to a homogeneous platoon, and the corresponding points distribution in the flow-density plot can be directly estimated by headway-velocity distributions or equivalently spacing-velocity distributions. As the aggregation time interval increases, the measured vehicles may belong to several separated homogeneous platoons that form heterogeneous ones. It becomes impossible to give a definite distribution for the corresponding points in the flow-density plot. But, most points for heterogeneous platoons still locate within a certain 2D region, whose boundaries can be obtained from homogeneous platoon model.

Chapter 6
Traffic Flow Breakdown Model Based on Headway/Spacing Distributions

6.1 Introduction

In the traditional sense of the transportation engineering, capacity is a deterministic and an inherent property of a road. It has been widely used as one of the basic concepts of transportation studies and plays a significant role in field applications. Numerous studies tend to treat road capacity as a static indicator for the guidance of road design and traffic control. According to "Highway Capacity Manual 2010" (Transportation Research Board of the National Academies 2010).

Definition 6.1 Capacity is the maximum sustainable hourly flow rate at which persons or vehicles reasonably can be expected to traverse a point or a uniform section of a lane or roadway during a given time period under prevailing roadway, environmental, traffic, and control conditions.

However, the road capacity is closely associated with external conditions, such as wet and dry road conditions, lighting conditions, road traffic management measures, etc. Although some scholars have noticed the random nature of the traffic capacity, and conducted useful discussions on the impact factors, overall, the stochastic concept of traffic capacity is still in a primary stage.

Recently, in modern traffic flow theory and applications, scholars broke shackle of the traditional assumption of deterministic capacity by investigating its randomness and time variability. Unlike the deterministic definition, such a dynamic and stochastic feature of capacity influences driving behaviors, travel time distribution, and the reliability of transportation systems.

At present, stochastic road capacity in bottlenecks of continuous traffic flow facilities has been widely accepted. The stochastic features of traffic capacity and the associated traffic breakdown phenomena become a hot research topic. Traffic breakdown occurs with persistent oscillations and leads to a wide range of spatial-temporal traffic congestions, so the studies of traffic breakdown are of great significance and beneficial to maintain the smoothness and efficiency of traffic flow.

© Tsinghua University Press, Beijing and Springer-Verlag Berlin Heidelberg 2015   117
X. (M.) Chen et al., *Stochastic Evolutions of Dynamic Traffic Flow*,
DOI 10.1007/978-3-662-44572-3_6

Definition 6.2 Traffic Flow Breakdown Probability (TF-BP) is the probability of traffic jams occurrence and upstream propagations when mathematical expectations of a particular traffic flow rate and the road capacity are given.

Based on the analysis of field measurements in Sect. 3.4, traffic breakdown phenomenon reflects the sudden transition between free-flow and congested states in the macroscopic perspective, spontaneously speed decreases in a relatively short time. Even in the same external conditions, observations of road capacity still show stochastic features (Chow et al. 2009), which means traffic breakdown may occur under different traffic demands. Road capacity distribution and TF-BP are closely related. And TF-BP directly reflects the reliability of traffic flow. This chapter establishes one kind of TF-BP model based on nonparametric statistics of survival data by using *Eulerian* data. Furthermore, we will build up the other kind of TF-BP model based on the headway/spacing distributions, then calibrate parameters and conduct simulations of phase diagrams by using *Lagrangian* data.

6.2 Nonparametric Lifetime Statistics Approach

Traffic flow breakdown is closely related to FD. By constructing the survival data of traffic capacity, we incorporate the approach of nonparametric statistics to the TF-BP problem. Under normal circumstances, TF-BP monotonically increases with the traffic flow rate. Weibull CDF is one of the most widely used curves to describe this relationship (Banks 2006; Brilon et al. 2005, 2001; Chow et al. 2009; Elefteriadou and Lertworawanich 2003; Elefteriadou et al. 1995, 2011; Geistefeldt and Brilon 2009; Kondyli and Elefteriadou 2011; Kondyli et al. 2011; Lorenz and Elefteriadou 2001; Mahnke and Kühne 2007; Ozbay and Ozguven 2007; Persaud et al. 2001; Shawky and Nakamura 2007; Shladover et al. 2010). For example, Brilon et al. (2005) studied the TF-BP model based on univariate Weibull distribution by using the right censored data. They took the traffic flow breakdown phenomenon as a failure event, estimated road capacity (survival time) by nonparametric statistical analysis, and then obtained the relationship between TF-BP and the upstream traffic flow rate.

Numerically, the curve of a univariate Weibull CDF is written as:

$$\mathcal{W}(q) = 1 - \exp\left[-\left(\frac{q}{\alpha}\right)^{\beta}\right] \tag{6.1}$$

where $\mathcal{W}(q)$ is the Weibull CDF, q is a random variable, α and β are the scale and shape parameters.

This model assumes TF-BP is only related with the upstream flow rate q_S, the TF-BP function and its PDF are as follows:

$$F_c(q_S) = \Pr\{c \leqslant q_S\} = W(q_S) \tag{6.2}$$

$$f_c(q_S) = \exp\left[-\left(\frac{q_S}{\beta}\right)^{\alpha}\right]\frac{\alpha}{\beta}\left(\frac{q_S}{\beta}\right)^{\alpha-1} \tag{6.3}$$

where $F_c(q_S)$ is the CDF of q_S or the lifetime distribution function, $1 - F_c(q_S)$ is the survival function, c is the stochastic road capacity, $f_c(q_S)$ is the derivative of $F_c(q_S)$ or the density function of the lifetime distribution (event density). Thus, the breakdown flow rate $\bar{q}_S$ is derived by

$$\bar{q}_S = \beta\Gamma\left(\frac{1}{\alpha}+1\right) \tag{6.4}$$

where $\Gamma(n) = \int_0^{\infty} \exp(-x)x^{n-1}dx$.

According to the Product Limit Method (PLM) proposed by Kaplan and Meier (1958), define the maximal likelihood function as:

$$L_{qs} = \prod_{i=1}^{n}[f_c(q_S^{(i)})]^{\theta_i}[1 - F_c(q_S^{(i)})]^{1-\theta_i} \tag{6.5}$$

where $q_S^{(i)}$ is the measured flow in the ith time interval, n is the number of measurements. If the mean measured speed in the ith time interval is larger than a specific threshold v^* but the value in the $i + 1$th time interval is smaller than v^*, then $\theta_i = 1$; otherwise, $\theta_i = 0$. Take the natural logarithm on both sides of Eq. (6.5), we have

$$\log L_{qs} = \sum_{i=1}^{n}\left\{\theta_i \log f_c(q_S^{(i)}) + (1 - \theta_i)\log[1 - F_c(q_S^{(i)})]\right\} \tag{6.6}$$

Since Kaplan–Meier's method is based on a generalized MLE (Kaplan and Meier 1958), the nonparametric MLE of the lifetime distribution function is

$$\widehat{F}_c(q_S) = 1 - \prod_{q_S^{(i)} \leqslant q_S} \frac{n_i - m_i}{n_i} \tag{6.7}$$

where $\ddot{F}_c(q_S)$ is the PLM estimate, n_i is the number of measured flows that satisfy $q_S^{(i)} \leqslant q_S$, m_i is the number of traffic congestions.

To maximize Eq. (6.5), let the partial derivatives of the likelihood function with respect to α and β equal 0, i.e.,

$$\frac{\partial \log L_{qs}}{\partial \alpha} = 0, \quad \frac{\partial \log L_{qs}}{\partial \beta} = 0 \tag{6.8}$$

The MLE of Weibull distribution parameters are obtained as follows:

$$\frac{1}{\hat{\alpha}} = \frac{\sum_{i=1}^{n} \log q_S^{(i)} [q_S^{(i)}]^{\hat{\alpha}}}{\sum_{i=1}^{n} [q_S^{(i)}]^{\hat{\alpha}}} - \frac{1}{r} \sum_{i=1}^{r} \log q_S^{(i:n)} \tag{6.9}$$

$$\hat{\beta} = \left\{ \frac{1}{r} \sum_{i=1}^{n} [q_S^{(i)}]^{\hat{\alpha}} \right\}^{1/\hat{\alpha}} \tag{6.10}$$

where $\hat{\alpha}$ is and implicit nonlinear equation that can be solved through numerical approximation, r is the number $\theta_i = 1$, satisfying $1 \leqslant r \leqslant n$, $\{q_S^{(1:n)}, q_S^{(2:n)}, q_S^{(r:n)}\}$ is the corresponding flow measurements.

On this basis, Chow et al. (2009) extended the univariate Weibull distribution to a bivariate Weibull distribution by taking into account the upstream traffic density ρ and mean speed v. For the statistical correlation of random variables ρ and v, their joint distribution is usually determined by the marginal distributions. At this point, the TF-BP can be formulated as a joint probability distribution function of density and mean speed, i.e.,

$$F_c(\rho, v) = C(\xi, \zeta) \tag{6.11}$$

where C is the Copula function that can be used to describe the dependence between random variables in probability theory and statistics, $\xi = \mathcal{W}(\rho|\alpha_\rho, \beta_\rho)$, $\zeta = \mathcal{W}(v|\alpha_v, \beta_v)$, and α_ρ, β_ρ, α_v, β_v are the parameter of Weibull distribution, respectively. Its PDF is

$$f_c(\rho, v) = \xi' \zeta' C'(\xi, \zeta) = \xi' \zeta' \frac{\partial^2 C(\xi, \zeta)}{\partial \xi \partial \zeta} \tag{6.12}$$

where $\xi' = f_c(\rho|\alpha_\rho, \beta_\rho)$, $\zeta' = f_c(v|\alpha_v, \beta_v)$.

Analogical to Eq. (6.5), define the maximal likelihood function as

$$L_{\rho,v} = \prod_{i=1}^{n} f_c(\rho_i, v_i) = \prod_{i=1}^{n} f_c(\rho_i) f_c(v_i) C'(\xi_i, \zeta_i) \tag{6.13}$$

let its partial derivatives with respect to α_ρ, β_ρ, α_v, β_v equal 0, then solve the MLE of $\hat{\alpha}_\rho$, $\hat{\beta}_\rho$, $\hat{\alpha}_v$, $\hat{\beta}_v$ by numerical methods.

This section briefly reviews the nonparametric statistical TF-BP model based on survival data. Road traffic flow is analogous to the survival time, and traffic breakdown is analogous to the failure event in survival analysis. By incorporating the product limit estimation of the survival function, we obtain the approximate Weibull relationship between TF-BP and the upstream traffic flow rate. However, the nonparametric statistics based on survival data only holds for the analysis of macroscopic measurements of traffic flow, such as road traffic flow, density, mean speed, etc. But it is difficult to dig more accurate information inherent in vehicle trajectories. In-depth analysis of the inherent correlation between microscopic vehicle trajectories and macroscopic traffic flow breakdown phenomena will be conducted in the following section.

6.3 Queueing Models for Breakdown Probability

6.3.1 Backgrounds

From the macroscopic point of view, traffic breakdown phenomena can be defined as a sudden transition from a free-flow state to a congested state. It is usually characterized by an obvious traffic velocity decrease within a short period of time. In order to maintain traffic fluency, the occurrence condition of traffic breakdown attracts continuous interests in the last decade (Bassan et al. 2006; Evans et al. 2001; Kerner and Klenov 2006; Smilowitz and Daganzo 2002).

As proven in many reports, traffic breakdown does not always occur at the same demand level, but can occur when flows are lower than the numerical value that is traditionally accepted as capacity (Banks 1991a, b; Brilon et al. 2005, 2001; Coifman and Kim 2011; Elefteriadou and Lertworawanich 2003; Elefteriadou et al. 1995; Geistefeldt and Brilon 2009; Hall and Agyemang–Duah 1991; Lorenz and Elefteriadou 2001; Persaud et al. 2001; Shawky and Nakamura 2007). So, many researchers and transportation engineers investigated the relationship between traffic breakdown probability and the upstream inflow rate, e.g., Coifman and Kim (2011), Shawky and Nakamura (2007).

Usually, this relationship is depicted as a monotonically increasing curve in the plot of breakdown probability versus inflow rate. The CDF for Weibull distribution is the most frequently incorporated function to fit the sigmoid breakdown probability curve. Weibull distribution was widely observed and verified according to empirical measurements from loop detectors (Banks 2006; Brilon et al. 2005, 2001; Chow et al. 2009; Elefteriadou and Lertworawanich 2003; Elefteriadou et al. 1995, 2011; Geistefeldt and Brilon 2009; Kondyli and Elefteriadou 2011; Kondyli et al. 2011; Lorenz and Elefteriadou 2001; Mahnke and Kühne 2007; Ozbay and Ozguven 2007; Persaud et al. 1998; Shawky and Nakamura 2007; Shladover et al. 2010).

Meanwhile, many researchers delved to seek dynamic explanations for these static breakdown probability curves in the last two decades. Some researchers discussed the traffic breakdown triggered by small perturbations and studied the associated queueing process (Bassan et al. 2006; Del Castillo 2001; Habib–Mattar et al. 2009; Jost and Nagel 2003; Kerner and Klenov 2006; Wang et al. 2007). In some recent works, traffic breakdown phenomena at an on-ramp bottleneck received increasing interests (Polus and Pollatschek 2002; Son et al. 2004). For example, Elefteriadou et al. (2011), Kondyli and Elefteriadou (2011), and Kondyli et al. (2011) discussed the relation between merging behaviors and traffic breakdown probability so as to reduce its occurrence by optimal ramp metering control strategies. Kühne and Lüdtke (2012) assumed the main road traffic flow was periodically perturbed by on-ramp vehicles. For each time, a jam queue (local congested vehicle cluster) will be generated after the merged vehicle. If this jam queue cannot dissipate before the next vehicle's merging, it often grows into a wide jam and we can observe a traffic breakdown finally. In other words, the traffic breakdown probability is assumed to be equivalent to the survival probability of the jam queue in a pre-selected time period. Via this way, we could link the macroscopic-level traffic breakdown probability at ramping region and the microscopic-level perturbation raised vehicle queueing process. This allows us to take advantage of modeling approaches in both sides.

Newell's simplified model (Newell 2002) provides a concise but an efficient way to formulate the jam queue formation, propagation, and dissipation. For example, Kim and Zhang (2008) developed a stochastic jam wave propagation model based on Newell's simplified car-following model to explain the growth and dissipation of a jam queue initialized by ramping vehicles. Based on their results, Son et al. (2004) further proposed a Monte-Carlo simulation method to study the survival probability of a jam queue. They assumed that the deceleration and acceleration waves caused by a ramp merging disturbance propagated upwards in a stochastic manner. In the simulation, a traffic breakdown was assumed to occur, if the deceleration wave caused by the previous merging vehicle was not eliminated when another consecutive merging disturbance happened. After thousands of times of simulations, the percent of noneliminated perturbations would be taken as the estimated traffic breakdown probability.

A merit of using Newell's simplified model (Newell 2002) in these models is to explain and simultaneously calculate the traffic breakdown probability according to empirical vehicular trajectories data (Kim and Zhang 2008; Son et al. 2004). However, the above models mainly focused on the jam wave propagation properties and did not thoroughly link the formation/dissipation of jam queue to traffic breakdown probability.

To solve this problem, we propose a new concise queueing model to interpret traffic breakdown mechanism. The merits of this queueing-based traffic breakdown model include:

(1) It provides a simple, intuitive yet insightful explanation for the S-shape traffic breakdown probability curve observed in practice: *the uncertainty of traffic breakdown mainly roots in the stochastic vehicle headways and the resulting joining times of the jam queue.*
(2) It further characterizes the distribution of the vehicle joining time in terms of upstream inflow rate, based on an extended Newell's car-following model. This enables us to directly calibrate model parameters via empirical measurements on normal traffic flow.
(3) It is a fast simulation model to estimate breakdown probability according to short-term measurements on headway/spacing distributions.

6.3.2 Some Previous Models

6.3.2.1 Jam Wave Model

In order to better describe the queueing model, let us recall some backgrounds associated with this study. To understand the complex behaviors of vehicular traffic flow, various car-following models have been proposed and analyzed. Kim and Zhang (2008) and Son et al. (2004) discussed how to explain and simulate traffic breakdown phenomena based upon Newell's simplified model. In their model, a perturbation was assumed to be caused by a merging vehicle at the ramping area.

Figure 6.1 illustrates the deceleration and acceleration waves that are triggered by this sudden interruption. The solid black curves stand for the main road vehicles, and the dash red curves stand for the merging vehicles. Suppose the loop detectors are deployed several meters in the upstream of the on/off-ramp bottleneck to measure the main road inflow rate, and they are plotted as a blue dash–dot line. If the deceleration and acceleration waves do not collide before the next vehicle merges from the ramp road, the jam queue will grow. Several jam queues will finally merge into a wide jam, and the wide jam will propagate backward and influence the traffic at the loop detector. From the collected data, we will observe a traffic breakdown occurs; see Fig. 6.1a.

On the contrary, Fig. 6.1b shows the dissipation of a jam queue, in which the deceleration and acceleration waves collide before the next merging vehicle enters. The jam region will not grow and the main road flow at the loop detector will not be influenced. Thus, we will not observe a breakdown from the traffic data collected at the loop detectors.

Kim and Zhang (2008) further studied whether a jam trigged by a merging vehicle could dissipate in a time period $\bar{H}$. Suppose the speed of the vehicle before it meets the deceleration wave is v_S, the speed of the vehicle after it meets the acceleration

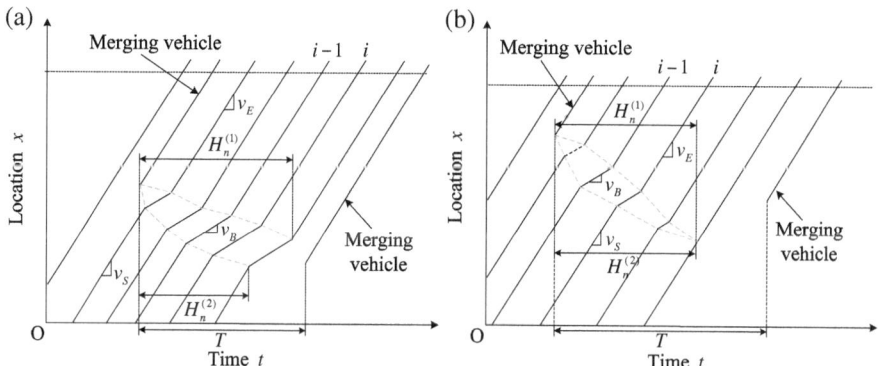

Fig. 6.1 The birth/death of jam queues from the viewpoint of Newell's simplified model. **a** Jam growth. **b** Jam dissipation

wave is $v_E = v_{\text{free}}$, while the speed of the vehicle within the jam queue is v_B. The wave propagation time from the merging vehicle to the nth vehicle in the acceleration and deceleration waves can be respectively written as:

$$
\left\{
\begin{aligned}
H_n^{(1)} &= \frac{1}{v_B - v_S}\left(v_B \sum_{i=1}^{n} \gamma_i^{B} - v_S \sum_{i=1}^{n} \gamma_i^{S} \right)\\
H_n^{(2)} &= \frac{1}{v_S - v_B}\left(v_E \sum_{i=1}^{n} \gamma_i^{E} - v_B \sum_{i=1}^{n} \gamma_i^{B} \right)
\end{aligned}
\right.
\tag{6.14}
$$

where $H_n^{(1)}$, $H_n^{(2)}$ are the wave propagation time of the two acceleration and deceleration waves, γ_i^{S}, γ_i^{B}, γ_i^{E} are the gap time for the ith following vehicle to travel the spacing minus the ith vehicle length before, within and after the deceleration wave, respectively. Due to the stochastic wave propagation properties, γ_i^{S}, γ_i^{B}, γ_i^{E} are random variables with given distributions.

In the ramp metering scenario, traffic breakdown is avoided when $H_n^{(1)} = H_n^{(2)}$. So the breakdown probability was defined by Son et al. (2004) as

$$
P_{\mathrm{B}} = 1 - \int_{0}^{+\infty} \Pr\{H_n^{(1)} = H_n^{(2)} \leqslant T\} p(v_B)\mathrm{d}v_B
\tag{6.15}
$$

where P_{B} is the breakdown probability, $p(v_B)$ is the probabilistic distribution of v_B.

Equation (6.15) links the macroscopic traffic breakdown phenomenon with the microscopic gap time variations. It provides us a chance to use the microscopic simulation to estimate traffic breakdown probability. The simulation results in Son et al. (2004) yielded an S-type curve in the breakdown probability versus inflow rate plot. This indicates that Newell's simplified model is capable of capturing the fundamental mechanism of traffic breakdown.

However, the breakdown probability curve obtained via this model is not smooth or steep enough, partly because of the insufficient modeling of gap time distribution. Therefore, we need to find a better model to fit the Weibull distribution-type breakdown probability that had been verified by many empirical experiments.

6.3.2.2 Random Walk Model

As shown in Fig. 6.2, when a vehicle joins the congested vehicle cluster, the jam queue increases in size; while, when a vehicle accelerates and leaves this jam queue, the jam queue decreases in size. Therefore, the rates of a vehicle that joins in or departs from the jam queue control the birth–death of it. If the joining rate is on average larger than the departing rate, a small perturbation will usually be amplified into a wide jam. Otherwise, a perturbation will finally dissipate.

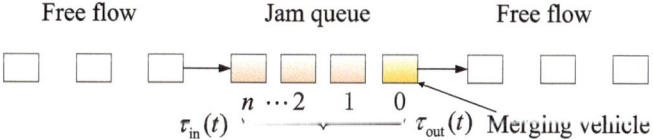

Fig. 6.2 The evolution of jam queue is controlled by both joining rate (*inflow*) and departing rate (*outflow*) of vehicles

These two competing processes were depicted as nucleation phenomena in Kühne et al. (2002), Mahnke and Kaupužs (2001), Mahnke et al. (2005), Mahnke and Kühne (2007), Mahnke and Pieret (1997), the growth of jam queue was modeled as a random walk, which was governed by the following discrete master equation

$$\frac{\partial \mathcal{P}(n,t)}{\partial t} = \omega_+(n-1)\mathcal{P}(n-1,t) + \omega_-(n+1)\mathcal{P}(n+1,t)$$
$$- [\omega_+(n)\mathcal{P}(n,t) + \omega_-(n)\mathcal{P}(n,t)] \tag{6.16}$$

where $\mathcal{P}(n,t)$ is the probability function when the jam queue length is equal to n at time t, $\omega_+(n)$ and $\omega_-(n)$ are the transition rates of the jam queue length increasing and decreasing from n, e.g., $\omega_+(n) = q_S$, $\omega_-(n) = 1/\tau_{out}$.

The departing rate was assumed to be a constant $2\,\mathrm{s}$ and the joining rate was described as a spacing-dependent car-following model. This one-dimensional random walk process Eq. (6.16) was first approximated by a continuous-time diffusion process and then solved by the decomposition of the corresponding Fokker–Planck equation in these literature studies. However, the solving algorithm is relatively complex.

6.3.3 G/G/1 Queueing Model

In this subsection, we derive the traffic breakdown probability for an on-ramp freeway bottleneck on the basis of queueing theory using microscopic vehicle trajectories. The model not only captures the dynamic evolution of queue size in the viewpoint of birth–death process, but also characterizes the equilibrium or steady features of jam queues. Hereafter, the merging area is assumed to be zero length for the sake of simplicity, so that on-ramp vehicles will enter the mainline only if the gap between two adjacent mainline vehicles is acceptable or large enough.

Different from models that focus on the jam wave propagation in Kim and Zhang (2008), Son et al. (2004), we characterize traffic breakdown merely in terms of jam queue dissipating time in this chapter. But other than the random walk model in Mahnke and Kaupužs (2001), we use the classical queueing theory to describe the evolution dynamics of the jam queue that is assumed to be first-in-first-out (FIFO) in this chapter. Moreover, we did not apply the hypothesized car-following model adopted in Mahnke and Kaupužs (2001) to derive the join time of the jam queue. Instead, we use the empirical measurements on headways to characterize the join time.

In short, the most innovative part is that we take vehicle headway as the prime factor to connect the upstream inflow rate and the joining time of the jam queue. Newell (2002) argued that the adjusting time was equal to the departing time and was not changeable with the velocity or the headway in front of a jam queue. However, we find vehicle headway changes with upstream inflow rate and thus influences the joining time. So, we extend the Newell's model to explain their relationships. We validate the mean vehicle headway is reciprocal to the upstream inflow rate. Moreover, the distributions of vehicle headways and the joining times are tightly related.

6.3.3.1 An Extension of Newell's Simplified Model

There were already several extensions of Newell's simplified model. One of these extensions assumes that $\tau_{in,i}$ (the adjusting time for the ith vehicle to pass the deceleration wave) are random variables due to the stochastic driving behaviors (Daganzo 2001; Kerner 2001; Smilowitz and Daganzo 2002; Yeo and Skabardonis 2009). We also assume $\tau_{in,i}$ to be random variables.

Since $\tau_{in,i}$ is tightly related with jam queue evolutions, we derive its statistic features by making the following three assumptions in the extend Newell's simplified model:

In the previous studies, (Chen et al. 2010a, b; Jin et al. 2009), headway/spacing between two consecutive vehicles were shown to follow a family of lognormal distributions. Based on this fact, we will further show that $\tau_{in,i}$ of the ith vehicle roughly follows a *lognormal* distribution.

(1) In accordance with our previous studies (Jin et al., 2009; Chen et al., 2010a, 2010b), vehicle headway/spacing are random variables that belong to shifted log-normal distributions with a constant scale parameter;
(2) Since we only consider the phase transition during traffic breakdown, the vehicles either run in free flow with velocity v_{free} or run in congested flow with velocity v_{cong};
(3) The jam propagation wave speed w is constant.

Suppose the random spacing s_i of the ith vehicle are independent random variables among drivers satisfying

$$s_i - s_0 \sim \text{Log-}\mathcal{N}\left(\mu_s, \sigma_s^2\right), \ i \in \mathbb{N}^+ \tag{6.17}$$

where s_0 is the infimum of spacing, μ_s and σ_s are the corresponding location and scale parameters, respectively.

Fig. 6.3   An illustration of
the geometric relationship
between $\tau_{\mathrm{in},i}$ and the spacing

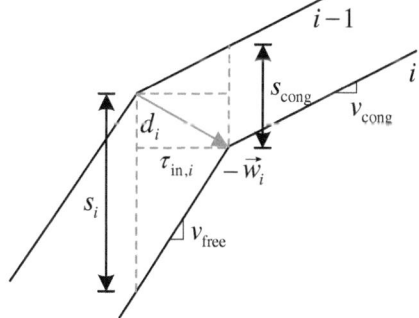

Since we only consider the phase transition during traffic breakdown, we can
assume the velocities before and after the jam wave front are two constants denoted
as $v_S = v_{\mathrm{free}}$ and $v_B = v_{\mathrm{cong}}$, respectively.

As shown in Fig. 6.3, we have the following relationship

$$d_i = s_i - v_{\mathrm{free}}\tau_{\mathrm{in},i} = s_{\mathrm{cong}} - v_{\mathrm{cong}}\tau_{\mathrm{in},i} \tag{6.18}$$

where s_{cong} is the compressed spacing in jam queue and is thus regarded to be
constant.

From Eq. (6.18), we have

$$\tau_{\mathrm{in},i} = \frac{s_i - s_{\mathrm{cong}}}{v_{\mathrm{free}} - v_{\mathrm{cong}}} = \theta_1 s_i + \theta_2 \tag{6.19}$$

where $\theta_1 = (v_{\mathrm{free}} - v_{\mathrm{cong}})^{-1} > 0$, $\theta_2 = -s_{\mathrm{cong}}(v_{\mathrm{free}} - v_{\mathrm{cong}})^{-1} < 0$ are the linear
coefficients.

Notice that $s_i \sim \mathrm{Log}\text{-}\mathcal{N}\left(\mu_s, \sigma_s^2\right) \Rightarrow \theta_1 s_i \sim \mathrm{Log}\text{-}\mathcal{N}\left(\mu_s + \log\theta_1, \sigma_s^2\right)$, so $\tau_{\mathrm{in},i}$
has a shifted lognormal distribution with the expectation and variance

$$\mathrm{E}[\tau_{\mathrm{in},i}] = \theta_1\mathrm{E}[s_i] + \theta_2$$
$$\mathrm{Var}[\tau_{\mathrm{in},i}] = \theta_1^2\mathrm{Var}[s_i]$$

In practice, v_{free} is much larger (Fig. 6.4) than v_{cong}, and s_{cong} is small. So θ_2
is usually very small. In order to calibrate the parameters in the shifted log-normal
distributions, we make the following assumption:

Proposition 6.1   *In general, the adjusting time τ_{in} for a vehicle to enter a jam queue
roughly follows a shifted log-normal distribution.*

$$\tau_{in} - \tau_0 \,\dot{\sim}\, \mathrm{Log}\text{-}\mathcal{N}\left(\mu_\tau, \sigma_\tau^2\right) \tag{6.20}$$

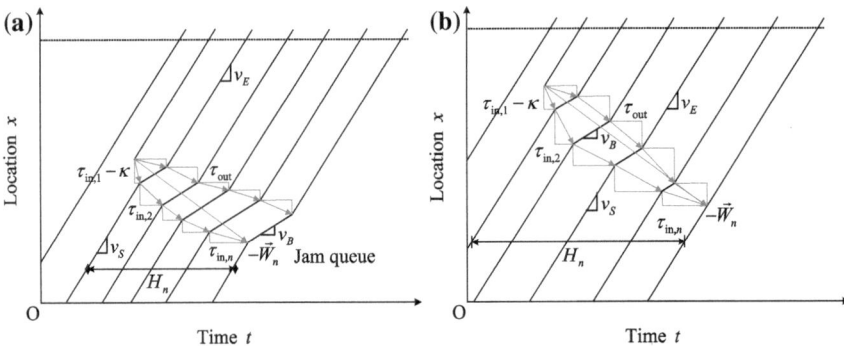

Fig. 6.4 The birth–death of the congested jam queue (vehicle cluster) from the viewpoint of the G/G/1 queueing model. **a** Jam growth. **b** Jam dissipation

where τ_0 is the infimum of adjusting time, μ_τ is the location parameter and σ_τ is the scale parameter of log-normal distribution.

The expectation (mean) and variance of τ_{in} are

$$E[\tau_{in}] = \exp(\mu_\tau + \sigma_\tau^2/2) + \tau_0$$
$$Var[\tau_{in}] = \exp(2\mu_\tau + \sigma_\tau^2)(\exp(\sigma_\tau^2) - 1)$$

Noticing that traffic breakdown probability is often measured versus upstream inflow rate, we need to know the distribution of τ_{in} with respect to the upstream inflow rate q_S. As shown in Fig. 6.5, the interarrival time of customers (vehicles) is independent and belongs to a general distribution. The remained job is to determine the influence of q_S on the lognormal distribution parameters μ_τ and σ_τ of τ_{in}.

The relationship between the expectation of spacing $E[s_i]$ and the corresponding inflow rate q_S can be written as:

$$q_S = \frac{v_{free}}{E[s_i]}, \quad \forall i \tag{6.21}$$

According to Eq. (6.18), the jam propagation wave speed of the ith vehicle is

$$w_i = \frac{d_i}{\tau_{in,i}} = \frac{s_i}{\tau_{in,i}} - v_{free} \tag{6.22}$$

then

$$s_i = (v_{free} + w_i)\tau_{in,i} \tag{6.23}$$

The deviation of an individual backward wave speed w_i to the average backward wave speed $w = E[w_i]$ is so small that we can omit the difference. This property can be illustrated as Fig. 6.5a, in which the white points in the left ellipse denote the traffic

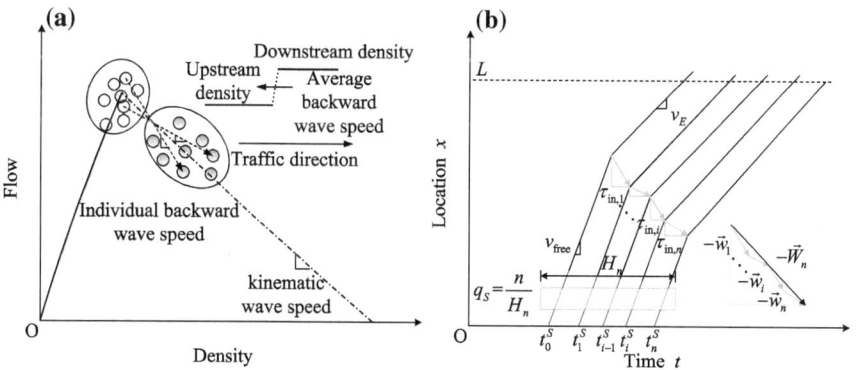

Fig. 6.5 a An illustration diagram of probabilistic expectation of backward wave speed; **b** An illustration of the relationship between q_S and the expectation of τ_{in}

flow states before breakdown, and the traffic flow rate after breakdown is denoted as the red points in the right ellipse. The average backward wave speed corresponds to the slope of the dash–dot line. Each individual backward speed w_i corresponds to a transition from a free-flow state (with low density) to a congested state (with high density), see the slopes of the dashed arrows in Fig. 6.5a. Generally, the traffic flow rate before any breakdown occurrence is generally not far away from maximum pre-breakdown inflow rate that was recorded. This feature guarantees that the slopes of those dashed arrows are roughly the same as the slop of the dash–dot line (or plainly, the average backward wave speed). So, we approximate w_i as $w_i = w$ in the rest of this chapter.

Suppose s_i and $\tau_{in,i}$ are independent random variables among drivers, get the expectations of the left and right terms of Eq. (6.23), we have

$$E[s_i] = (v_{free} + w)E[\tau_{in}] \tag{6.24}$$

Substitute Eq. (6.24) into Eq. (6.21), we have the relationship between the traffic flow rate q_S and $E[\tau_{in}]$ as

$$q_S = \frac{v_{free}}{E[\tau_{in}](v_{free} + w)} \tag{6.25}$$

Alternatively, Eq. (6.25) can be derived from some different approaches. As shown in Fig. 6.5b, the relationship between the upstream inflow rate q_S and the mean adjusting time $\tau_{in,i}$ of the inflow can be written as:

$$q_S = E\left[n \left(\sum_{i=1}^{n} \tau_{in,i} \left(1 + \frac{\|-\mathbf{W}_n\|}{v_S} \right) \right)^{-1} \right] = \frac{v_S}{E[\tau_{in}](v_S + w)} \tag{6.26}$$

where $-\mathbf{W}_n$ is the jam wave propagation velocity, i.e., $-\mathbf{W}_n = -\sum_{i=1}^{n} \mathbf{w}_i$. Its scalar value w satisfies

$$w = \frac{\sum_{i=1}^{n} \|-\mathbf{w}_i\|_x}{\sum_{i=1}^{n} \|-\mathbf{w}_i\|_t} = \frac{\|-\mathbf{W}_n\|_x}{\|-\mathbf{W}_n\|_t} = E[w_i] \tag{6.27}$$

The derived formulas of q_S and w are similar to the formulas that had been obtained based upon the framework of the stochastic wave propagation model (Kim and Zhang 2008) and the passing rate measurement (Chiabaut et al. 2009).

We can conclude that μ_τ decreases with the inflow rate q_S

$$\mu_\tau = \log\left(\frac{v_{\text{free}}}{q_S(v_{\text{free}} + w)} - \tau_0\right) - \frac{\sigma_\tau^2}{2} \tag{6.28}$$

Moreover, we also assume that σ_τ is a constant. This assumption is partly verified via field data. Specifically, in order to capture both w_i and d_i of the ith vehicle from microscopic vehicle trajectories, Duret et al. (2008) proposed a method to calculate the trajectory shift in the space-time coordinate by minimizing the distinction between simulated and experimental vehicles' trajectories. Following this idea, Chiabaut et al. (2010) estimated the jam wave velocity as:

$$\|-w\| = \arg\min_u \varepsilon(\mathbf{u}_{|x}(\bar{v}, u)) = \arg\min_u \frac{\sqrt{\frac{1}{K}\sum_{k=1}^{K}\left(\mathbf{u}_{|x}(\bar{v}_k, u) - \overline{\mathbf{u}_{|x}(\bar{v}_k, u)}\right)^2}}{\overline{\mathbf{u}_{|x}(\bar{v}_k, u)}}$$
$$\tag{6.29}$$

where $\varepsilon(\cdot)$ means the relative standard deviation, $-\mathbf{w}$ is the congestion wave speed vector, whose projection at space coordinate is the transition distance denoted as $\mathbf{w}_{|x} \equiv d$, and at time coordinate is the reaction time for transition denoted as $\mathbf{w}_{|t} \equiv \tau$, respectively. $\|-\mathbf{w}\|$ is the scalar of $-\mathbf{w}$, satisfying $\|-\mathbf{w}\|_x = d$, $\mathbf{u}_{|x}(\bar{v}_i, u)$ is the projection at space coordinate of the ith velocity range within n ranges for the passing velocity vector $\mathbf{u}$, $\mathbf{u}_{|x}(\bar{v}_i, u)$ is the average passing velocity for all velocity ranges.

Similarly, we can estimate τ_{in} from vehicle trajectories as:

$$\tau_{\text{in}} = \arg\min_{\mathbf{u}} \varepsilon(\mathbf{u}_{|t}(\bar{v}, u)) = \arg\min_{\mathbf{u}} \frac{\sqrt{\frac{1}{K}\sum_{k=1}^{K}\left(\mathbf{u}_{|t}(\bar{v}_k, u) - \overline{\mathbf{u}_{|t}(\bar{v}_k, u)}\right)^2}}{\overline{\mathbf{u}_{|t}(\bar{v}_k, u)}} \tag{6.30}$$

In order to better investigate the optimal value of τ_{in} for every two consecutive trajectories, we only check the consecutive vehicles in a stable platoon (each vehicle remains driving in the same lane without lane changing and the spacing between two consecutive vehicles were smaller than 50 m). Table 6.1 and Fig. 6.6 show the statistics of the entering inflow rates of three 15 min intervals of the Highway 101 NGSIM dataset. The five lanes mean inflow rate slightly decreased with time from 1,626 to 1,414 veh/h. The observed τ_{in} include 1,272 samples, with a minimum observation

Table 6.1 Validation of the shifted lognormal τ_{in} distribution using NGSIM dataset

Time period	7:50–8:05	8:05–8:20	8:20–8:35	Total dataset
Entering vehicles (Lane 1–5)	2032	1887	1768	5687
Average inflow (veh/h/lane)	1626	1510	1414	1521
Sampling trajectories	339	469	464	1272
v_S (m/s)	10.3	9.3	8.6	–
$\hat{\mu}_\tau$	−0.032	−0.023	−0.051	−0.036
$\hat{\sigma}_\tau$	0.457	0.436	0.448	0.446
$E[\tau_{in}]$	1.469	1.470	1.444	1.461
$Var[\tau_{in}]$	0.223	0.208	0.201	0.214
p-value	0.257	0.303	0.217	0.040
K-S statistics	0.055	0.045	0.049	0.039
Cutoff value	0.088	0.075	0.075	0.046
Accept the hypothesis	Yes	Yes	Yes	Yes

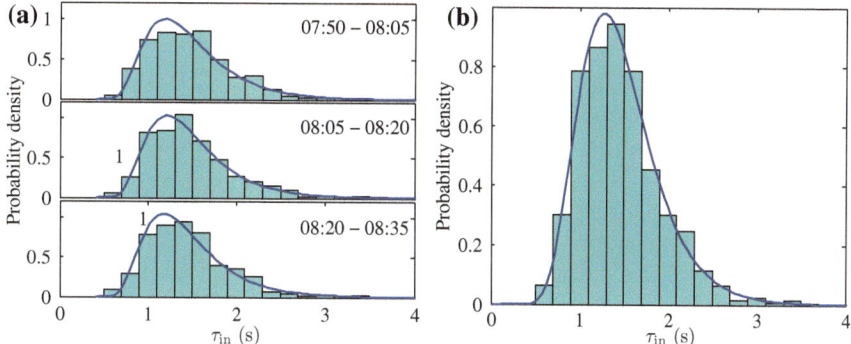

Fig. 6.6 Lognormal fitting of the adjusting time distribution of the Highway 101 NGSIM dataset. **a** 15 min intervals. **b** Total 45 min

of 0.544 s. To identify shifted lognormal distribution, we choose the infimum of adjusting time to be $\tau_0 = 0.4$ s. The calibrated values of $\hat{\mu}_\tau = E[\log(\tau_{in} - \tau_0)]$ is negative because the mean of the shifted log-normally distributed random variable $\tau_{in} - \tau_0$ is mainly within $(0, 1]$. The Kolmogorov–Smirnov (KS) test results for the distribution of the total dataset at the significance level of $\alpha = 0.01$ show the p value is 0.040, the test statistic is 0.039, and the cutoff value for determining whether the test statistic is significant is 0.046, which is larger than the test statistic. Therefore, we accept the shifted lognormal distribution of τ_{in}. It is clear that σ_τ almost keeps a constant. This partly verifies our assumption that σ_τ is a constant.

In the following, we set $\sigma_\tau = 0.446$. As suggested in many literature studies (e.g., Kim and Zhang 2008), we set $w \approx 18$ km/h $= 5$ m/s and $v_{free} = 20$ m/s in the following. For a given q_S, we can directly obtain the corresponding μ_τ and also the PDF of $\tau_{in,i}$, see Fig. 6.7.

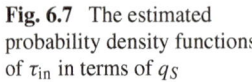
Fig. 6.7 The estimated
probability density functions
of τ_{in} in terms of q_S

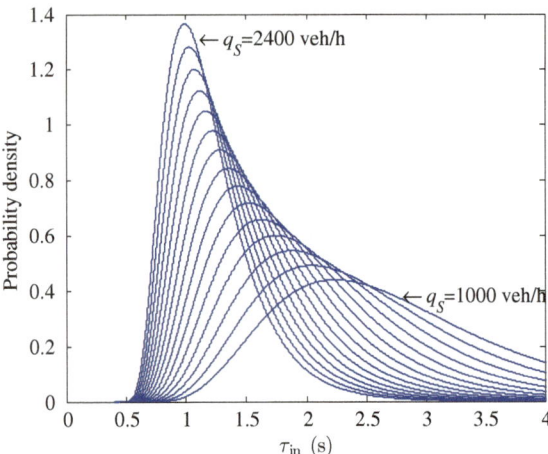

6.3.3.2 G/G/1 Queueing Model Derivation

In our model, the time for a vehicle to join this jam queue is taken as the interarrival
time τ_{in} of customers (vehicles). On the other side, we regard the time for a vehicle
to depart from this jam queue as the service time τ_{out} (the adjusting time for a vehicle
to pass the acceleration wave). $\{\tau_{in,i},\ 1 \leq i \leq n\}$ should be a series of independently and identically shifted lognormal distributed random variables. Similar to the
assumptions in Mahnke and Pieret (1997), we first assume τ_{out} is a constant. This
assumption infers that the jam queue can be described by a G/D/1 queueing model.

According to Elefteriadou et al. (2011), Kondyli and Elefteriadou (2011), Kondyli
et al. (2011), Mahnke et al. (2005), the breakdown probability can be influenced by
a variety of factors, including the geometry of ramping regions, the multilane effect
and possible overtaking maneuvers. In this simplified model, we assume all these
factors can be reflected as different pre-selected values of τ_{out}. After entering the
queue, customers (vehicles) are patient and willing to wait for infinite time before
leaving this queue because overtaking behaviors are forbidden.

Clearly, the waiting buffer of this queue is also infinite, if we assume the highway
is long enough. It is worth noticing that the vehicles in the jam queue are not static but
might move forward at a small velocity. But, we only care about the size evolution
of this queue but not its spatial evolution.

In summary, according to Kendall's shorthand notation, we model this queue as
a G/G/1 queue:

(1) With a given inflow rate q_S, the arrival intervals are independent and identi-
    cally distributed shifted lognormal type random numbers (General distribution
    of interarrival time, G);

(2) General distribution of service time or the departing time $\tau_{out,i}$ for the ith vehicle to pass the acceleration wave (General distribution of inter-departure time, G). Since people perform various time delays to depart from a jam queue in reality, i.e., stochastic response time during acceleration. To check the possible influence of a stochastic τ_{out}, we could assume that $\tau_{out,i}$ is a random variable whose mean value keeps constant. For simplicity, we assume $\{\tau_{out,1}, \ldots, \tau_{out,n}\}$ to be i.i.d. Gaussian random variables in this chapter as

$$\tau_{out} = c + \varepsilon \tag{6.31}$$

where c is a constant that denotes the default value of departing time, $\varepsilon \sim \mathcal{N}(0, 0.2^2)$ is the Gaussian white noise.

(3) Only one server or single lane.

(4) Stochastic extra delay of the first vehicle. The duration of the merging disturbance is stochastic and associated with many vehicle dynamic factors, e.g., merging vehicle velocity, accepted gap. We could consider all these factors simultaneously by taking the assumption that κ is also a random variable. Because we do not have enough a priori knowledge on its real distribution, without loss of generality, we assume κ is under Gaussian type white disturbances

$$\kappa = 0.5 + \mathcal{N}(0, 0.2^2) \tag{6.32}$$

The initial size of this queue is one vehicle (the first vehicle influenced by the merging vehicle). Then, the survival process of this jam queue can be depicted by a set of mutual and independent probabilistic events $\{\mathcal{A}_i, \ i \in \mathbb{N}^+\}$, satisfying $\bigcap_{i=1}^{+\infty} \mathcal{A}_i = \emptyset$ and $\bigcup_{i=1}^{+\infty} \mathcal{A}_i = \Omega$, where $\mathcal{A}_i$ denotes that the ith vehicle leaves the jam queue.

We can characterize traffic breakdown mainly in terms of jam queue dissipating time. More precisely, we have

(1) The probability that this queue dissipates, when the first vehicle leaves (i.e., in 2 s), is

$$P_1 = \Pr\{\mathcal{A}_1\} = \Pr\{\tau_{in}(1) - \kappa \geqslant \tau_{out}\} \tag{6.33}$$

where the random variable is $\tau_{in}(1) = \tau_{in,1}$, κ is the disturbance delay of the first vehicle. Since the first vehicle interrupted by the merging disturbance has a relatively shorter time to adjust its velocity, we introduce an extra delay κ. As a result, the breakdown probability will become larger, and we discuss the sensitivity of the breakdown probability to κ.

(2) The probability that this queue dissipates, right when the second vehicle leaves (in other words, in 4 s), is

$$P_2 = \Pr\{\mathcal{A}_2\} = (1 - P_1)\Pr\{\tau_{in}(2) - \kappa \geqslant 2\tau_{out} | \tau_{in}(1) - \kappa < \tau_{out}\} \tag{6.34}$$

where the random variable is $\tau_{in}(2) = \tau_{in,1} + \tau_{in,2}$. The first part $1 - P_1$ means Scenario 2) $\mathcal{A}_2$ happens only if Scenario (1) $\mathcal{A}_1$ does not happen.

(3) Similarly, for $i \geqslant 2$, the probability that this queue dissipates, right when the ith vehicle leaves, is

$$P_i = \Pr\{\mathcal{A}_i\}$$

$$= \left(1 - \sum_{j=1}^{i} P_j\right) \Pr\{\tau_{\text{in}}(i) - \kappa \geqslant i\tau_{\text{out}}|$$

$$\tau_{\text{in}}(1) - \kappa < \tau_{\text{out}}, \ldots, \tau_{\text{in}}(i-1) - \kappa < (i-1)\tau_{\text{out}}\} \qquad (6.35)$$

where the random variable is $\tau_{\text{in}}(i) = \sum_{j=1}^{i} \tau_{\text{in},j}$.

Thus, for a large n, the truncated approximation of the breakdown probability is

$$P_{\text{B}} = P_{+\infty} \approx 1 - \sum_{i=1}^{n} P_i = \bigcap_{m=1}^{n} \mathcal{A}_m^C$$

$$= \Pr\left\{\mathcal{A}_m^C | \mathcal{A}_{m-1}^C, \ldots, \mathcal{A}_1^C\right\}$$

$$= \Pr\{\tau_{\text{in}}(1) < \tau_{\text{out}} + \kappa, \ldots, \tau_{\text{in}}(n) < n\tau_{\text{out}} + \kappa\} \qquad (6.36)$$

where $\mathcal{A}_m^C$, $m = 1, \ldots, n$ is the complementary set of $\mathcal{A}_m$, $\tau_{\text{in}}(n) = \sum_{i=1}^{n} \tau_{\text{in},i}$ is the summation of the adjusting time of n vehicles.

The conditional probability in Eq. (6.36) leads to complicated convolution calculations and is thus difficult to analytically solve. The upper and lower bounds of P_{B} can be derived as follows:

The PDF of τ_{in} is

$$f_\tau(\tau_{\text{in}}) = \frac{1}{\sqrt{2\pi}\sigma_\tau(\tau_{\text{in}} - \tau_0)} \exp\left(-\frac{(\log(\tau_{\text{in}} - \tau_0) - \mu_\tau)^2}{2\sigma_\tau^2}\right) \qquad (6.37)$$

The CDF is

$$\mathcal{F}(\tau_{\text{in}}) = \Phi\left(\frac{\log(\tau_{\text{in}} - \tau_0) - \mu_\tau}{\sigma_\tau}\right) \qquad (6.38)$$

where $\Phi(\cdot)$ is the standard normal distribution function.

Although there is no closed form of distribution for a sum of i.i.d. lognormal random variables, the summation of the characteristic time approximately belongs to the lognormal type distribution (Beaulieu and Rajwani 2004; Beaulieu and Xie 2004; Nadarajah 2008; Romeo et al. 2003), i.e.,

$$\tau_{\text{in}}(m) - m\tau_0 \stackrel{\cdot}{\sim} \text{Log-}\mathcal{N}\left(\mu_{\tau(m)}, \sigma_{\tau(m)}^2\right), \quad 1 \leqslant m \leqslant n \qquad (6.39)$$

where $\tau_{\text{in}}(m) = \sum_{i=1}^{m} \tau_{\text{in},i}$. Thus, the CDF of $\tau_{\text{in}}(m) - m\tau_0$ is

$$\mathcal{F}_m(\tau_{\text{in}}(m)) \approx \Phi\left(\frac{\log(\tau_{\text{in}}(m) - m\tau_0) - \mu_{\tau(m)}}{\sigma_{\tau(m)}}\right) \tag{6.40}$$

where the parameters are

$$\begin{cases} \sigma_{\tau(m)}^2 = \log[\exp(\sigma_\tau^2) + m - 1] - \log m \\ \mu_{\tau(m)} = \mu_\tau + (\sigma_\tau^2 - \sigma_{\tau(m)}^2)/2 + \log m \end{cases} \tag{6.41}$$

and the expectation and variance of $\tau_{\text{in}}(m)$ is

$$\begin{cases} \mathrm{E}[\tau_{\text{in}}(m)] = \exp(\mu_{\tau(m)} + \sigma_{\tau(m)}^2/2) + m\tau_0 = m\exp(\mu_\tau + \sigma_\tau^2/2) + m\tau_0 = m\mathrm{E}[\tau_{\text{in}}] \\ \mathrm{Var}[\tau_{\text{in}}(m)] = \exp(2\mu_{\tau(m)} + \sigma_{\tau(m)}^2)(\exp(\sigma_{\tau(m)}^2) - 1) = m^2\mathrm{Var}[\tau_{\text{in}}] \end{cases} \tag{6.42}$$

One lower bound of P_B is

$$P_B \geqslant \Pr\{\tau_{\text{in}}(n) < n\tau_{\text{out}} + \kappa\}$$

$$+ \Pr\left\{\bigcap_{m=1}^{n-1} (\tau_{\text{in}}(m) < m\tau_{\text{out}} + \kappa)\right\} - 1$$

$$\geqslant \sum_{m=1}^{n} \Pr\{\tau_{\text{in}}(m) < m\tau_{\text{out}} + \kappa\} - n + 1$$

$$\approx \sum_{m=1}^{n} \mathcal{F}_m(m\tau_{\text{out}} + \kappa) - n + 1$$

$$= \sum_{m=1}^{n} \Phi\left(\frac{\log(m(\tau_{\text{out}} - \tau_0) + \kappa) - \mu_{\tau(m)}}{\sigma_{\tau(m)}}\right) - n + 1 \tag{6.43}$$

At the same time, another lower bound of P_B is

$$P_B \geqslant \Pr\{\tau_{\text{in},1} < \tau_{\text{out}} + \kappa\} \prod_{i=2}^{n} \Pr\{\tau_{\text{in},i} < \tau_{\text{out}}\}$$

$$= \mathcal{F}(\tau_{\text{out}} + \kappa)[\mathcal{F}(\tau_{\text{out}})]^{n-1}$$

$$= \Phi\left(\frac{\log(\tau_{\text{out}} - \tau_0 + \kappa) - \mu_\tau}{\sigma_\tau}\right) \times \left[\Phi\left(\frac{\log(\tau_{\text{out}} - \tau_0) - \mu_\tau}{\sigma_\tau}\right)\right]^{n-1} \tag{6.44}$$

Therefore, the lower bound of P_B is the maximum of the right terms in Eqs. (6.43) and (6.44).

On the other hand, one upper bound of P_B is

$$
\begin{aligned}
P_{\mathrm{B}} &\leqslant \min_{1\leqslant m\leqslant n} \left\{ \Pr\left\{ \tau_{\mathrm{in}}(m) < m\tau_{\mathrm{out}} + \kappa \right\} \right\} \\
&\approx \min_{1\leqslant m\leqslant n} \left\{ \mathcal{F}_m \left(m\tau_{\mathrm{out}} + \kappa \right) \right\} \\
&= \min_{1\leqslant m\leqslant n} \left\{ \Phi\left(\frac{\log\left(m(\tau_{\mathrm{out}} - \tau_0) + \kappa \right) - \mu_{\tau(m)}}{\sigma_{\tau(m)}} \right) \right\}
\end{aligned}
\tag{6.45}
$$

It is worthy to note that the equalities in the formulated lower and upper bounds of P_{B} can be satisfied if and only if the events $\{\tau_{\mathrm{in}}(m) < m\tau_{\mathrm{out}} + \kappa, \ m = 1, \ldots, n\}$ are independent. However, these estimated bounds are not so tight in general.

6.3.3.3 Monte-Carlo Simulation Procedure and Results

The conditional probabilities in Eq. (6.36) consist of complicated convolution calculations and are thus difficult to analytically solve. In order to calculate the accurate traffic breakdown probability, we need to resort to Monte-Carlo simulations (Kalos and Whitlock 2008). More precisely, we will simulate to check whether a newly formed jam queue could dissipate in a given time period H_n, e.g., $H_n = 1$ min.

As illustrated in Fig. 6.5b, the number of approaching vehicles within a time period H_n is about $\hat{n} = \lfloor q_S H_n \rfloor$, where $\lfloor z \rfloor$ denotes the largest integer not greater than z. We do not need to simulate for a long-time period H_n that is larger than 60 s, because $\Pr\{\tau_{\mathrm{in}}(n) < n\tau_{\mathrm{out}} + \kappa\} \to 1$ for a large enough n. This further reduces our simulation time.

The basic simulation outflow is shown in Fig. 6.8, which includes the following steps:

Step 0: Given the upstream inflow q_S, τ_{out} and a long enough time window H_n. Initialize the number of traffic breakdown $B = 0$.

Step 1: Generate $\bar{n}$ i.i.d. random variables $\{\tau_{\mathrm{in},1}, \ldots, \tau_{\mathrm{in},\hat{n}}\}$ that follow the log-normal distribution with $\bar{\mu}_\tau$, $\bar{\sigma}_\tau^2$. Check whether the breakdown criteria in Eq. (6.36) is satisfied, if yes, let $B = B + 1$; if no, go to Step 2.

Step 2: Repeat Step 1 for 10,000 times.

Step 3: Calculate the simulated breakdown probability as $P_{\mathrm{B}}(q_S) = B/10000$. Choose another value of q_S and repeat the whole procedure until we obtain the entire traffic breakdown curve.

In the following simulation, we increase q_S from 1,000 to 2,500 veh/h with a step of 50 veh/h to get the traffic breakdown probability curve. We designed several experiments to test the proposed model. Constrained by the length limit, we only present two most important experiments in this book.

In the first experiment, we examine the possible influence of departing time on traffic breakdown probability. Particularly, we are interested whether the proposed queueing model performs significantly different when we adopt the constant

Fig. 6.8 Estimation approach
by Monte-Carlo method

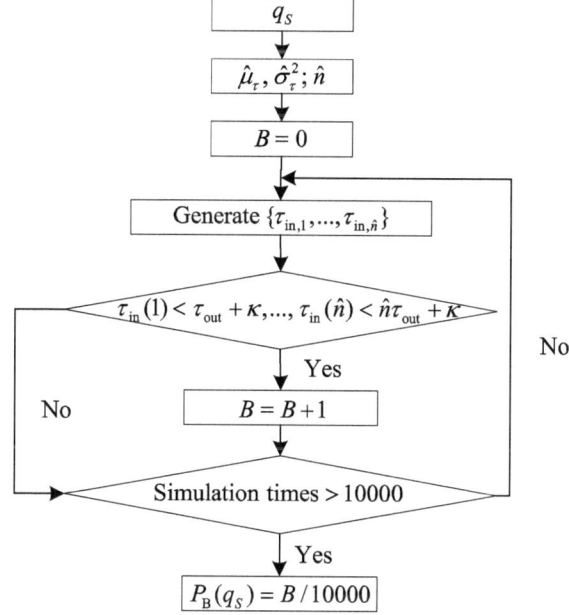

Fig. 6.9 Comparisons
between the G/G/1 queueing
model and the fitting Weibull
distributions in terms of τ_{out},
given $(\kappa = 0.5)\,\text{s}$

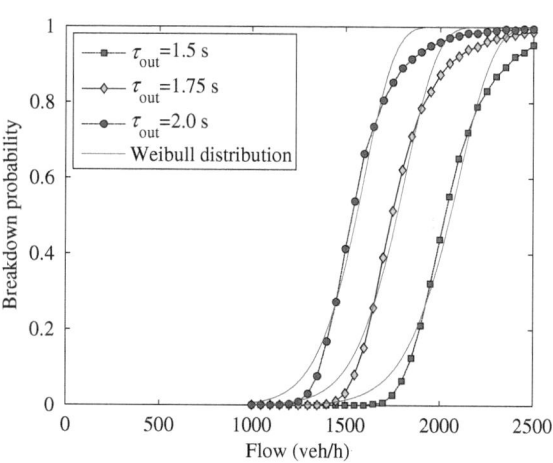

departing time as Mahnke and Kaupužs (2001), Kühne et al. (2002) instead of the
stochastic departing time.

Figure 6.9 shows the simulation results of traffic breakdown probability curves
with different τ_{out} varying from 1.5 to 2.0 s, given $\kappa = 0.5\,\text{s}$. All of these curves
show nondecreasing S-style features, which are consistent with many other studies
(Lorenz and Elefteriadou 2001; Son et al. 2004; Wang et al. 2010a).

In order to compare the simulation results with the widely used Weibull CDF curve, we use the least-squares criteria to obtain the Weibull CDF curve that best fits the simulated traffic breakdown probability curve

$$R(\alpha, \ \beta) = \sum_{qs} (\mathcal{W}(q_s|\alpha, \ \beta) - P_B(q_S))^2 \qquad (6.46)$$

where $R(\alpha, \ \beta)$ is the square residuals between the breakdown probability curve $P_B(q_S)$ and the Weibull distribution with parameters of α, β. The fitting parameters can be gotten as:

$$\hat{\alpha}, \ \hat{\beta} = \arg\min_{\alpha, \ \beta} R(\alpha, \ \beta) \qquad (6.47)$$

then, the least-squares residual is

$$\text{LSR} = \sum_{qs} \left(\mathcal{W}(q_s|\hat{\alpha}, \ \hat{\beta}) - P_B(q_S) \right)^2 \qquad (6.48)$$

We set $v_{\text{free}} = 20\,\text{m/s}$, $w = 5\,\text{m/s}$, $H_n = 60\,\text{s}$ in the simulations. Test results show that the curves obtained by the G/G/1 queueing model are not as steep as the Weibull CDF-type breakdown probability curves, especially when q_S approaches the maximum pre-breakdown inflow rate that had been recorded. Moreover, the traffic breakdown probability curve moves rightward when τ_{out} decreases. In particular, when τ_{out} is reduced by 0.1 s, the mean road capacity will increase by around 6 %.

Table 6.2 compares the fitting parameters for Weibull distributions shown in Fig. 6.9 by using the least-squares residual. As the increase of τ_{out} in the G/D/1 queueing model, both the estimated scale parameter $\hat{\alpha}$ and shape parameter $\hat{\beta}$ decrease, and the estimated Weibull distributions move leftward.

Figure 6.10 shows the sensitivity analysis results of κ, when $\tau_{\text{out}} = 2.0\,\text{s}$ and other parameters are equal to Fig. 6.9. Generally, the larger the extra delay κ is, the larger the breakdown probability will be. As illustrated in Fig. 6.10a, a larger κ will notably shift the breakdown probability curve upper-left and make it more steep. Figure 6.10b shows the LSR values almost decrease with κ for all τ_{out} values. It indicates that when we choose a larger value of κ, the simulation curve will better fit with the CDF curve of Weibull distribution.

Table 6.2 Comparisons of the optimal Weibull distributions between G/D/1 queueing models, given $\kappa = 0.5\,\text{s}$

τ_{out}	$\hat{\alpha}$	$\hat{\beta}$	LSR/10^{-3}
1.5	2127.3	11.9	47.3
1.6	2009.9	11.0	52.7
1.7	1902.0	10.5	52.5
1.8	1810.2	10.0	52.6
1.9	1724.8	9.5	53.9
2.0	1646.9	9.2	51.5

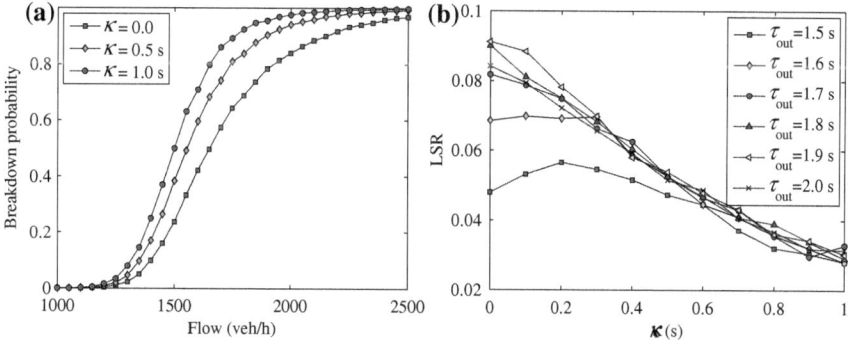

Fig. 6.10 Parameter sensitivity of κ, given ($\tau_{out} = 2.0$ s). **a** Breakdown probability curves. **b** Estimated LSR of Weibull distributions

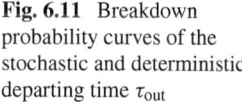

Fig. 6.11 Breakdown probability curves of the stochastic and deterministic departing time τ_{out}

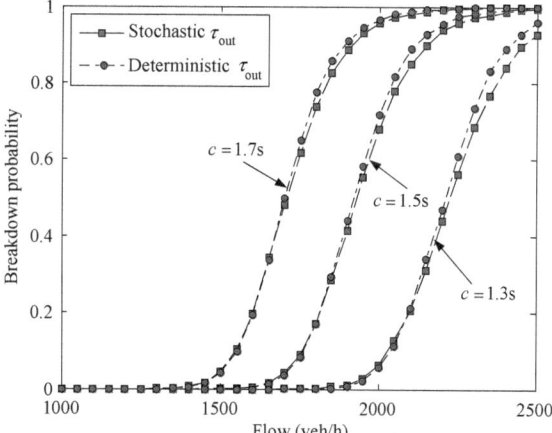

Figure 6.11 shows that the differences of these two kinds of breakdown probability curves are quite small. The queueing model yields nondecreasing sigmoid breakdown probability curves and captures dominant factors that govern the traffic breakdown dynamics around an on-ramp bottleneck, which are consistent with many other studies (Lorenz and Elefteriadou 2001; Son et al. 2004; Wang et al. 2010b). The stochastic τ_{out} curves are usually a little lower than the deterministic curves. Moreover, the traffic breakdown probability curve moves rightward when $E[\tau_{out}]$ decreases.

In the second experiment, we investigate the possible impact of additive disturbance κ on traffic breakdown probability. Figure 6.12 shows that $E[\kappa]$ dominates the shape of breakdown probability curve. Generally, the larger $E[\kappa]$ is, the larger breakdown probability will be. Moreover, $E[\kappa]$ notably shifts the breakdown probability curve upper-left and makes it much steeper.

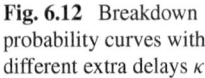

Fig. 6.12 Breakdown probability curves with different extra delays κ

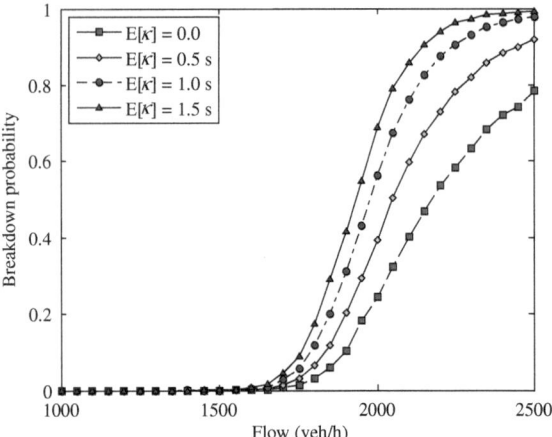

6.3.4 Discussions

Now, we can come back to the initial question: *what on earth leads to the Weibull type breakdown probability curve?* Our answer is the dominating factor should be the average headway that changes with the main road flow rather than the shape of vehicle headway distribution.

Based on Newell's model, the average headway will decrease as the inflow rate increases, which leads to a longer existing time of the jam queue and thus a larger probability of breakdown. So, traffic breakdown probability is sensitive to the relation between the average jam queue joining time τ_{in} and the jam queue departing time τ_{out}. As a result, we could observe a sharp increase of breakdown probability around the critical flow rate, where $\tau_{in} = \tau_{out}$.

Further simulations indicate that we could observe a similar (but slightly different) breakdown probability curve, if we use another one-peak, smooth, and right skewed distribution model with the same mean value and variance; because the dominating factor should be the average headway that changes with the main road flow.

We will categorize three kinds of jam queues associated with ramping vehicle disturbances by using a simplified temporal-spatial queueing model. The jam queue width will decrease if $\tau_{in} > \tau_{out}$, keep constant if $\tau_{in} = \tau_{out}$, and increase if $\tau_{in} < \tau_{out}$. Further simulations indicate that we could observe a similar (but slightly different) breakdown probability curve, if we use other one-peak, smooth, and right skewed distribution models instead of the shifted log-normal distribution model and keep the same mean value and variance; because the dominating factor should be the average headway that changes with the main road flow, and the PDF shapes will not influence the breakdown probability curve too much.

Besides, we may sometimes observe the so-called pinned localized clusters, which stay at a fixed location over a longer period of time but do not grow into severe congestions (Helbing 2009; Schönhof and Helbing 2007; Treiber et al. 2000, 2010).

Usually, traffic velocity and throughput will not be significantly influenced when vehicles pass pinned localized clusters or moving localized clusters. Since the formed jam queue did not dissipate in a given time, the proposed model will mistakenly treat such phenomena as traffic breakdowns. However, the existing condition of such phenomena is relatively strict so that the occurring probability of such phenomena is quite small. Neglecting such phenomena in our simulations will not notably change the shape of traffic breakdown probability curve.

6.3.5 Model Validation

To validate the breakdown probability estimation algorithm, we compare the simulated breakdown probability curves with the field measurements of breakdown events.

We first extract the three-month (June 1 to August 31, 2005), 5-min interval, lane-by-lane vehicle counts, speeds, and occupancies from the loop detectors at Station 717490 on the southbound Highway 101, California, via PeMS (2011). This is the same site where the trajectory data of NGSIM 101 Highway were collected. The NGSIM trajectory data used here were collected on Wednesday, June 15, 2005. We use the average values across five lanes to calculate the traffic breakdown probability curve. Figure 6.13 shows the empirical relationships between flow rates and speed (or occupancy). The critical speed and occupancy that correspond to the maximum flow rate (i.e., road capacity) are approximately 88.5 km/h and 15 %, respectively.

In this book, we define the so-called pre-breakdown flow rate estimated from time occupancies similarly to Lorenz and Elefteriadou (2001): if the occupancy at time t is smaller than the critical occupancy, i.e., $O(t) < O_c$, and $O(t + 1)$, $O(t+2)$, $O(t+3) > O_c$, then the flow rate at time t, i.e., $q(t)$, is the pre-breakdown flow rate. Figure 6.14 shows an example of one-day time series of flow rates and time occupancies, in which two breakdown events were observed. The breakdowns occurred at 7:15 am in the morning rush hour and 5:45 pm in the afternoon rush hour, respectively.

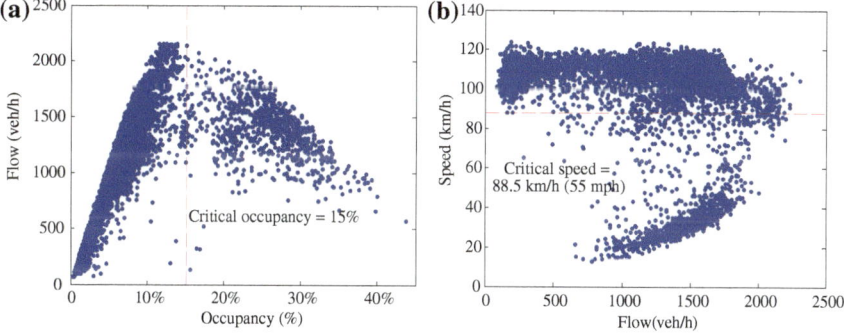

Fig. 6.13 Empirical measurements of loop detectors. **a** Flow rate and time occupancy. **b** Flow rate and speed

Fig. 6.14 Characteristics of breakdown flow rates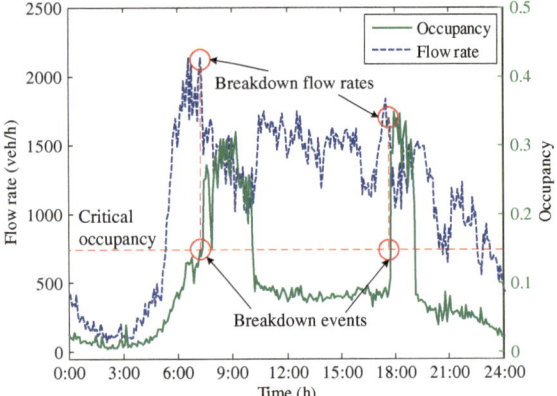

Fig. 6.15 Comparison of the breakdown probabilities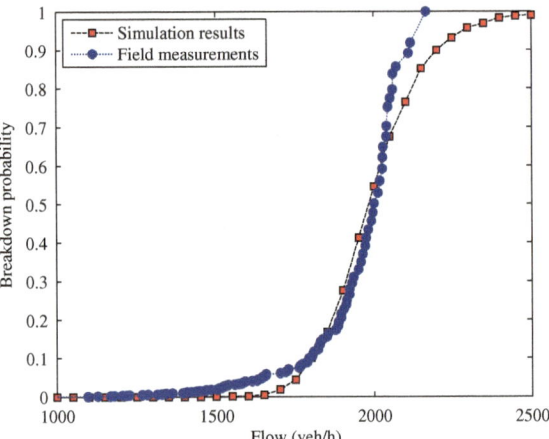

The corresponding breakdown flow rates are 2,131 and 1,646 veh/h, respectively. Totally, we observed 118 breakdown events from the three-month field measurements. The minimum and maximum breakdown flow rates are 835 and 2,166 veh/h, respectively.

The PLM adopted from Brilon et al. (2005) was applied in this study. Figure 6.15 shows the empirical and simulated breakdown probabilities. The parameters in the G/G/1 model were calibrated from NGSIM dataset as: $\hat{\mu}_\tau = -0.036$, $\hat{\sigma}_\tau = 0.446$, $E[\tau_{in}] = 1.461$ s, $\tau_0 = 0.4$ s. Since we do not have a priori knowledge on the expectations of τ_{out} and κ, they are assumed to be equal to the mean value of τ_{in}, i.e., $E[\tau_{out}] = E[\kappa] = E[\tau_{in}] = 1.461$ s.

The comparison shows that the predicted breakdown probability estimations could fit well with field measurements by loop detectors. These curves yield an analogous tendency and curvature, which validate the proposed queueing-based breakdown probability estimation method in this chapter. However, we also find that the simulated curve is a little lower than the empirical curve since 2,000 veh/h. This is mainly because the proposed model does not thoroughly describe the "self-compressions" phenomena happened in practice (Helbing 2009; Schönhof and Helbing 2007). Indeed, when the upstream inflow rate becomes larger than a threshold (i.e., 2,000 veh/h), the newly developing small jam queues will soon merge into a large jam queue whose dissipating speed is much slower than that of the small jam queues.

6.3.6 Summary

In this section, we study the evolution process of a jam queue initialized by an on-ramp vehicle and calculate the resulting traffic breakdown probability from a microscopic viewpoint using vehicle trajectories. Different from many previous studies that discuss the relationship between jam wave propagation and breakdown probability, we address on the size evolution of the jam queue instead of its spatial evolution.

Following this idea, a simple G/G/1 queueing model is proposed based on Newell's simplified car-following model. We show that the growth and dissipation of a jam queue are controlled by two competing stochastic processes: the upstream vehicle joining process and downstream vehicle departing process. The average joining rates are governed by the upstream inflow rate, while the departing rates are assumed to be stochastic (we incorporate the normal distribution to departing time). With the increase of the inflow rate, the average joining rate grows significantly but the departing rate is constrained, which therefore yields the sharply increase of traffic breakdown probability when upstream inflow rates approach the capacity flow rate. Monte-Carlo simulation examples show that the simulated breakdown probability curves agree with empirical breakdown probability curve.

The main merit of the proposed intuitive model in this chapter is that it links the microscopic measurement (headways/spacings of an individual vehicle or platoons) and macroscopic measurement (traffic breakdown phenomena, dynamics of jam formation, propagation, and dissipation).

Although a practical model of breakdown probability should be more complex to include weather, surface condition, and sight quality, et al., we believe this queueing model is useful and could be further applied in many applications, including designing optimal ramp metering strategies (Kesting et al. 2010; Kondyli and Elefteriadou 2011) and cooperative driving strategies.

6.4 Phase Diagram Analysis

6.4.1 Backgrounds

Congestions caused by on-ramp flow remain as an important research topic during the last 50 years. In order to describe the formation, evolution, and dissipation of such congestions, various models had been proposed (Ahn et al. 2010; Cassidy et al. 2002; Chen et al. 2010b; Daganzo et al. 1999; Duret et al. 2010; Helbing et al. 2009; Kerner 2009, 2004; Laval and Leclercq 2010b; Leclercq et al. 2011; Lu et al. 2010; Munjal and Pipes 1971; Schönhof and Helbing 2007; Treiber et al. 2000, 2010; Windover and Cassidy 2001; Zhang and Shen 2009). Particularly, the "phase diagram" approach received increasing interests in the last decade, since it provided a tool to categorize the rich congestion patterns and explain the origins of some complex phenomena.

The term of "phase" is borrowed from thermodynamics into traffic flow theory to denote different spatial-temporal states (patterns) of traffic flows. Usually, we enumerate the possible conditions and plot all the observed traffic states into one diagram, from which we can semi-quantitatively determine the phase transition conditions.

One application of phase diagram is to sketch the boundaries of different congestion patterns observed in ramping regions on freeways. If the ramping flow rate exceeds a certain limit, we observe two kinds of congestions: spatially localized and spatially extended congestions. In the first kind of congestions, the jam queue of vehicles will be restricted within a certain low-velocity region whose length is roughly stable over time. Differently, in the second kind of congestions, the lengths and locations of low-velocity regions may change over time. We can further classify these congestions into several categories, e.g., moving jam queues and stop-and-go waves, etc. By examining the conditions and boundaries of these traffic congestion patterns, we can gain more insights into traffic flow dynamics.

In order to reproduce and then explain the empirical phase diagram observed in practice, some microscopic car-following and lane-changing models had been proposed and discussed in the last two decades (Chen et al. 2010b; Helbing et al. 2009; Kerner 2009; Lu et al. 2010; Schönhof and Helbing 2007; Treiber et al. 2000, 2010). However, most of these models incorporated quite a lot of parameters and were thus relatively too complicated to calibrate or analyze quantitatively. This problem hinders us from discovering the governing power that characterizes the mechanism of phase formation and transition.

To solve this problem, we propose a spatial-temporal queueing model to depict traffic jams via the minimum number of parameters and rules. The new model applies a specialized Newell's simplified model (Newell 2002) to describe the trajectories of vehicles. As a result, we can link individual vehicle movements and the resulting congestion patterns. This helps reveal the factors that dominate the formations and transitions of different congestion patterns.

To present a detailed discussion on these new findings, a new queueing model for traffic congestions is presented based on the extended Newell's simplified model. This study further discusses how to apply this queueing model to obtain the analytical phase diagram for a typical ramping region and provides simulation results to support the derivation.

6.4.2 The Spatial-Temporal Queueing Model

Newell's simplified model (Ahn et al. 2004; Kim and Zhang 2008; Newell 2002; Son et al. 2004; Yeo 2008; Yeo and Skabardonis 2009) characterizes how a vehicle changes its speed according to its leading vehicle. When the vehicles run on a homogeneous highway, this model assumes that the time-space trajectory of the following vehicle is the same as the leading vehicle, except for a transformation in both space and time.

Based on Newell's simplified model, Kim and Zhang (2008) analyzed the growth and dissipation of jam queue caused by perturbations. It was shown that stochastic reaction times of drivers resulted in stochastic ac/deceleration wave speeds and thus led to random growth/decay of disturbances.

Differently in Mahnke and Kaupužs (2001), Mahnke et al. (2005), Mahnke and Kühne (2007), Mahnke and Pieret (1997), the temporal evolution instead of the spatial evolution of a jam queue was addressed. The evolution is characterized by how a vehicle joins in and departs from it. Thus, we can depict the evolution dynamics of the jam queue as a diffusion process as well.

Similarly, we can describe the jam queue as a FIFO queue system and then depict its temporal evolution by queueing theory. But, if we want to simultaneously describe the spatial evolution features of congestion patterns, we need a new queueing model that also considers the spatial evolution of a jam queue.

In the following, let us consider a specialized Newell's model that focuses on the scenario of congestion propagations. For the sake of simplicity, we assume that there exist only two kinds of vehicle velocities: free-flow velocity v_{free} and congested velocity v_{cong}. Therefore, we only have two kinds of slopes for all the piecewise-linear trajectories shown in Fig. 6.16a. Obviously, the birth–death of a jam queue is controlled by the rate of vehicles that join in/depart from it. If the joining rate is on average larger than the departing rate, a small perturbation will grow into a wide jam. Otherwise, the perturbation will dissipate finally. As shown in Fig. 6.16a, we define the joining time and departing time for the ith vehicle as $\tau_{\text{in},i}$ and $\tau_{\text{out},i}$, respectively. In the rest of this section, we will show that the congestion patterns are indeed determined by $\tau_{\text{in},i}$ and $\tau_{\text{out},i}$ through an implicit way.

Moreover, we assume the propagation velocity of upstream shock front (congestion wave) as w, i.e., the slope of the dashed arrow in Fig. 6.16a; when the leading $(i-1)$th vehicle joins a jam queue, the following ith vehicle will adjust its velocity after a spatial displacement of

$$d_i = w\tau_{\text{in},i} \tag{6.49}$$

According to the geometric relationship shown in Fig. 6.16a, the development of spacing between two vehicles can be written as:

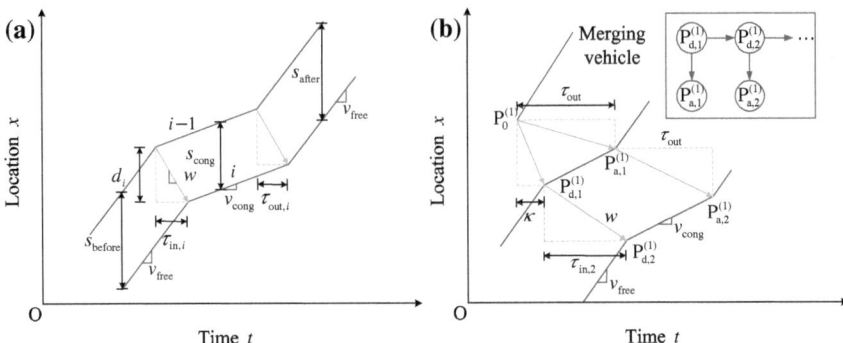

Fig. 6.16 Calculation of spatial-temporal queueing model. **a** An illustration of the Newell's simplified. **b** Determination of ac/deceleration points

$$
\begin{cases}
s_{\text{before}} = (v_{\text{free}} + w)\tau_{\text{in},i} \\
s_{\text{cong}} = (v_{\text{cong}} + w)\tau_{\text{in},i} \\
s_{\text{after}} = (v_{\text{cong}} + w)\tau_{\text{in},i} + (v_{\text{free}} - v_{\text{cong}})\tau_{\text{out}}
\end{cases}
\tag{6.50}
$$

where s_{before} is the spacing before the ith vehicle joining the jam queue, s_{cong} is the spacing within the jam queue, and s_{after} is the spacing after the ith vehicle departing from the jam queue.

Moreover, we can denote the corresponding headway as:

$$
h_{\text{before}} = s_{\text{before}}/v_{\text{free}} = (v_{\text{free}} + w)\tau_{\text{in},i}/v_{\text{free}}
\tag{6.51}
$$

Therefore, the change of spacing through a jam queue is

$$
s_{\text{after}} - s_{\text{before}} = (v_{\text{free}} - v_{\text{cong}})(\tau_{\text{out}} - \tau_{\text{in},i})
\tag{6.52}
$$

If $\tau_{\text{in},i} < \tau_{\text{out}}$, the spacing becomes larger after the jam queue; if $\tau_{\text{in},i} = \tau_{\text{out}}$, the spacing maintains constant, which is the assumption in Newell's simplified model (Newell 2002), if $\tau_{\text{in},i} > \tau_{\text{out}}$, the spacing becomes smaller after the jam queue.

As illustrated in Fig. 6.16b, in order to get the spatial evolution of a jam queue, we only need to determine the speed changing points of each vehicle and then link these points via a series of lines segments with known slopes. When the first vehicle is determined, we will take a two-step recursive method to get the trajectories of the following vehicles: first determine the time axis of a speed changing point and then get its associated spatial axis via the following geometric assumptions.

In the first jam queue, suppose the first ramping vehicle's merging point is $P_0^{(1)}(t_0^{(1)}, x_0^{(1)})$, where the superscript "(1)" denotes the first jam queue and the subscript "0" represents the merging vehicle. Denote the first deceleration point of the first influenced vehicle on main road as $P_{d,1}^{(1)}(t_{d,1}^{(1)}, x_{d,1}^{(1)})$ in the plot. If the deceleration wave of the first vehicle influences the speed of its following vehicle, we can determine the first deceleration point $P_{d,2}^{(1)}(t_{d,2}^{(1)}, x_{d,2}^{(1)})$ of the second vehicle by

$$
\begin{cases}
t_{d,2}^{(1)} = t_{d,1}^{(1)} + \tau_{in,2}^{(1)} \\
x_{d,2}^{(1)} = x_{d,1}^{(1)} - w\tau_{in,2}^{(1)}
\end{cases}
\tag{6.53}
$$

where $\tau_{in,2}^{(1)}$ is the join time of the second vehicle in the first jam queue.

This determination process of deceleration points can be formulated as the following recursion

$$
\begin{cases}
t_{d,i}^{(1)} = t_{d,i-1}^{(1)} + \tau_{in,i}^{(1)} \\
x_{d,i}^{(1)} = x_{d,i-1}^{(1)} - w\tau_{in,i}^{(1)}
\end{cases}
\quad i = 2, \ldots s, n
\tag{6.54}
$$

Suppose this propagation process in the first jam queue is denoted as $P_{d,1}^{(1)} \rightarrow P_{d,2}^{(1)}$. Similarly, we can sequentially determine the deceleration points for the vehicles coming next, i.e., $P_{d,i}^{(1)} \rightarrow P_{a,i}^{(1)}$, $i = 2, 3 \ldots$.

As suggested in Mahnke and Kaupužs (2001), Mahnke et al. (2005), Mahnke and Kühne (2007), Mahnke and Pieret (1997), we assume a constant departing time of τ_{out} for any vehicle. So, we can determine the acceleration point $P_{a,1}^{(1)}(t_{a,1}^{(1)}, x_{a,1}^{(1)})$ of the first influenced vehicle as

$$
\begin{cases}
t_{a,1}^{(1)} = t_{d,1}^{(1)} + \tau_{out} \\
x_{a,1}^{(1)} = x_{d,1}^{(1)} + v_{cong}\tau_{out}
\end{cases}
\tag{6.55}
$$

As shown in Fig. 6.16b, the acceleration point $P_{a,2}^{(1)}(t_{a,2}^{(1)}, x_{a,2}^{(1)})$ of the second vehicle should be

$$
\begin{cases}
t_{a,2}^{(1)} = t_{d,2}^{(1)} + 2\tau_{out} - \tau_{in,2}^{(1)} \\
x_{a,2}^{(1)} = x_{d,2}^{(1)} + v_{cong}(\tau_{out} - \tau_{in,2}^{(1)})
\end{cases}
\tag{6.56}
$$

unless the acceleration wave collide the deceleration wave.

Similarly, we could sequentially determine the acceleration points for the vehicles coming next, i.e., $P_{d,i}^{(1)} \rightarrow P_{a,i}^{(1)}$ as:

$$
\begin{cases}
t_{a,i}^{(1)} = t_{d,i}^{(1)} + i\tau_{out} - \displaystyle\sum_{j=2}^{i} \tau_{in,j}^{(1)} \\
x_{a,i}^{(1)} = x_{d,i}^{(1)} + v_{cong}\left(i\tau_{out} - \displaystyle\sum_{j=2}^{i} \tau_{in,j}^{(1)}\right)
\end{cases}
\quad i = 2, \ldots, n
\tag{6.57}
$$

On the other hand, we determine the de/acceleration points of the ith trajectory based on the $(i-1)$th trajectory in the kth jam queue, i.e., $P_{d,i-1}^{(k)} \rightarrow P_{d,i}^{(k)}$ and $P_{a,i-1}^{(k)} \rightarrow P_{a,i}^{(k)}$ as:

$$\mathbf{Y}_i^{(k)} = \mathbf{Y}_{i-1}^{(k)} + \mathbf{b}_i^{(k)}, \quad i = 2, \dots, n \tag{6.58}$$

where

$$\mathbf{Y}_i^{(k)} = \begin{bmatrix} t_{d,i}^{(k)} \\ x_{d,i}^{(k)} \\ t_{a,i}^{(k)} \\ x_{a,i}^{(k)} \end{bmatrix}, \quad \mathbf{b}_i^{(k)} = \begin{bmatrix} \tau_{in,i}^{(k)} \\ -w\tau_{in,i}^{(k)} \\ \tau_{out} \\ v_{cong}\tau_{out} - (v_{cong}+w)\tau_{in,i}^{(k)} \end{bmatrix} \tag{6.59}$$

Sometimes, two jam queues collide and finally merge into one; see Fig. 6.17. Suppose the ith vehicle in the first jam queue is the same as the jth vehicle in the second jam queue. We have $i > j$, because the first jam queue occurs earlier than the second one. Then, the merging condition of two jam queues is

$$t_{a,i}^{(1)} \geqslant t_{d,j}^{(2)} \quad \text{or} \quad x_{a,i}^{(1)} \geqslant x_{d,j}^{(2)} \tag{6.60}$$

In case this happens, we let $t_{d,j}^{(2)} = t_{a,i}^{(1)}$ and $x_{d,j}^{(2)} = x_{a,i}^{(1)}$ in order to avoid driving back or unreasonable trajectories, see Fig. 6.17b.

Figure 6.18 shows the trajectory of a vehicle that is influenced by multiple jam queues. The join time of this vehicle in the kth jam queue is denoted as $\tau_{in}^{(k)}$. We can see that $\tau_{in}^{(k)}$ changes with the number of jam queues

$$\begin{cases} s_{cong}^{(1)} = s_{before}^{(1)} - (v_{free} - v_{cong})\tau_{in}^{(1)} \\ s_{after}^{(1)} = s_{before}^{(2)} = s_{cong}^{(1)} + (v_{free} - v_{cong})\tau_{out} \end{cases} \tag{6.61}$$

If the vehicle joins another jam queue after departing from the previous one, we can calculate the join time to the second jam queue by

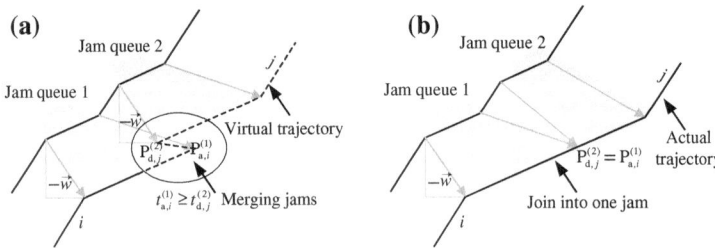

Fig. 6.17 An illustration of how to simulate two colliding jam queues. **a** The unrevised trajectory. **b** The revised trajectory

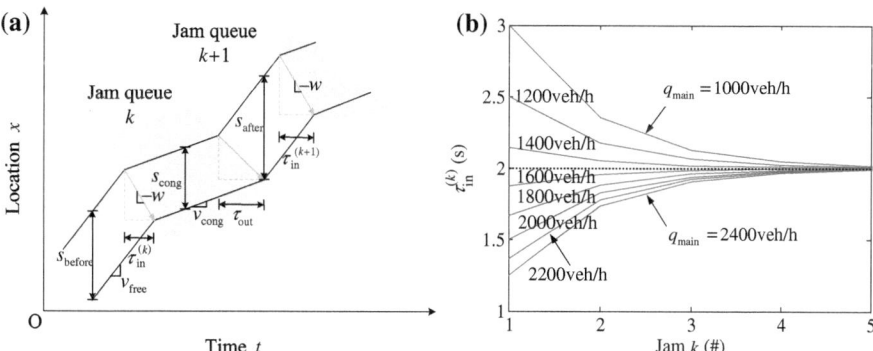

Fig. 6.18 Situation of passing multiple jam queues. **a** An illustration of changing join times. **b** Asymptotic properties of $\tau_{in}^{(k)} \to \tau_{out}$

$$\tau_{in}^{(2)} = \frac{s_{after}^{(1)}}{v_{free} + w} = \frac{(v_{cong} + w)\tau_{in}^{(1)} + (v_{free} - v_{cong})\tau_{out}}{v_{free} + w} \qquad (6.62)$$

Extending the logic into a general situation, we have the recursion formula for the dynamic evolution of join time

$$\tau_{in}^{(k+1)} = \frac{(v_{cong} + w)\tau_{in}^{(k)} + (v_{free} - v_{cong})\tau_{out}}{v_{free} + w} = a\tau_{in}^{(k)} + b, \; k = 1, 2\cdots \quad (6.63)$$

where $a = (v_{cong} + w)/(v_{free} + w) < 1$ and $b = (1 - a)\tau_{out}$.

Therefore, the general term formula of $\tau_{in}^{(k)}$ is

$$\tau_{in}^{(k)} = a^{k-1}\tau_{in}^{(1)} + (1 - a^{k-1})\tau_{out}, \; k = 1, 2\cdots \qquad (6.64)$$

For a large k, the limit of $\tau_{in}^{(k)}$ is equal to τ_{out}, i.e.,

$$\lim_{k \to \infty} \tau_{in}^{(k)} = \tau_{out} \qquad (6.65)$$

We can also obtain the asymptotic properties of $\tau_{in}^{(k)}$ as:

$$\begin{cases} \tau_{in}^{(1)} < \tau_{out} \Rightarrow \tau_{in}^{(k)} < \tau_{in}^{(k+1)} < \tau_{out} \\ \tau_{in}^{(1)} = \tau_{out} \Rightarrow \tau_{in}^{(k)} = \tau_{in}^{(k+1)} = \tau_{out} \\ \tau_{in}^{(1)} > \tau_{out} \Rightarrow \tau_{out} < \tau_{in}^{(k+1)} < \tau_{in}^{(k)} \end{cases}, \; k = 1, 2 \cdots \qquad (6.66)$$

$s_{cong}^{(k)}$ is smaller than both $s_{before}^{(k)}$ and $s_{after}^{(k)}$. So, we only need to check whether $s_{cong}^{(k)} \geqslant s_{safe}$ holds in the simulations. We have

$$s_{\text{cong}}^{(k)} = s_{\text{cong}}^{(1)} - (v_{\text{free}} - v_{\text{cong}}) \sum_{i=2}^{k} \tau_{\text{in}}^{(i)} + (k - 1)(v_{\text{free}} - v_{\text{cong}}) \tau_{\text{out}}$$

$$= s_{\text{cong}}^{(1)} - (v_{\text{free}} - v_{\text{cong}})(\tau_{\text{in}}^{(1)} - \tau_{\text{out}}) \sum_{i=2}^{k} a^{i-1} \tag{6.67}$$

Clearly, we have the limit of $s_{\text{cong}}^{(k)}$ as

$$\lim_{k \to \infty} s_{\text{cong}}^{(k)} = s_{\text{cong}}^{(1)} - (v_{\text{free}} - v_{\text{cong}})(\tau_{\text{in}}^{(1)} - \tau_{\text{out}}) \frac{a}{1 - a}$$

$$= s_{\text{cong}}^{(1)} - (v_{\text{cong}} + w)(\tau_{\text{in}}^{(1)} - \tau_{\text{out}})$$

$$= (v_{\text{cong}} + w)\tau_{\text{out}} > s_{\text{safe}} \tag{6.68}$$

Based on Eq. (6.67), we can see the series of $s_{\text{cong}}^{(k)}$ is monotonous with k and $\lim_{k \to \infty} s_{\text{cong}}^{(k)} > s_{\text{safe}}$. So, we only need to check whether we have $s_{\text{cong}}^{(1)} > s_{\text{safe}}$.

On the other hand, our model needs to preserve the convective instability Treiber and Kesting (2011). This means small oscillations grow due to merging disturbances, but the range of growing amplitudes propagate only upstream. In the extreme case that the acceleration wave does not propagates upstream, the situation satisfies the following relationship:

$$\tau_{\text{in}} w = (\tau_{\text{out}} - \tau_{\text{in}}) v_{\text{cong}} \tag{6.69}$$

The condition that ensures convective instability is

$$\frac{\tau_{\text{in}}^{(k)}}{\tau_{\text{out}}} > \frac{v_{\text{cong}}}{w + v_{\text{cong}}}, \quad k = 1, 2 \ldots \tag{6.70}$$

According to the asymptotic property of the join time denoted in Eq. (6.66), we have three situations:

(1) If $\tau_{\text{in}}^{(1)} < \tau_{\text{out}}$, then $\tau_{\text{in}}^{(k)} < \tau_{\text{in}}^{(k+1)} < \tau_{\text{out}}$, we need to select $\tau_{\text{in}}^{(1)} > \tau_{\text{out}} v_{\text{cong}} (w + v_{\text{cong}})^{-1}$;

(2) If $\tau_{\text{in}}^{(1)} = \tau_{\text{out}}$, the left term of Eq. (6.70) equals 1, while the right term is less than 1, so Eq. (6.70) can be naturally satisfied;

(3) If $\tau_{\text{in}}^{(1)} > \tau_{\text{out}}$, the left term of Eq. (6.70) is larger than 1 and is also larger than the right term, so Eq. (6.70) can also be naturally satisfied. Therefore, our model guarantees the convective instability if $\min\{\tau_{\text{in}}^{(k)}\} > \tau_{\text{out}} v_{\text{cong}} (w + v_{\text{cong}})^{-1}$, $k = 1, 2 \ldots$.

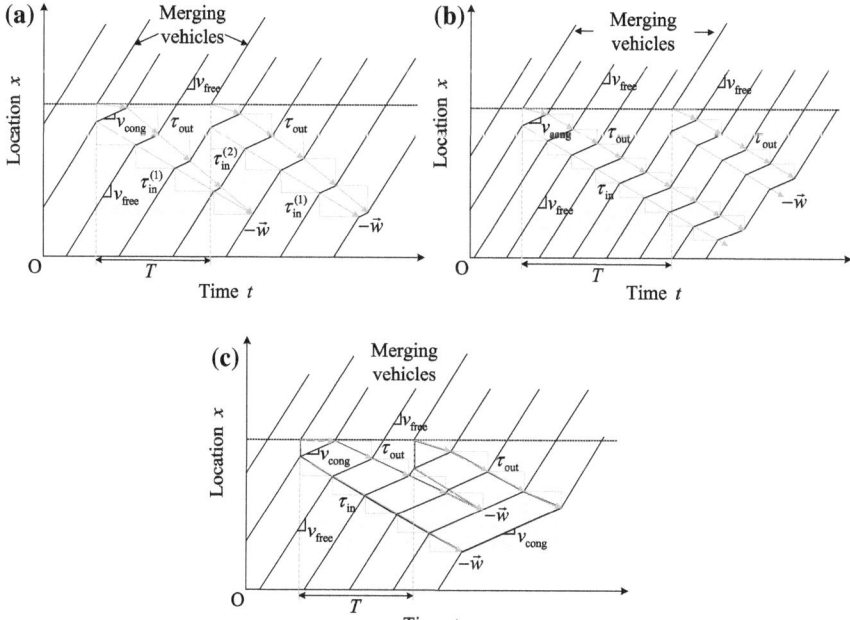

Fig. 6.19 Three typical jam queues from the viewpoint of queueing theory. **a** Jam queue dissipation. **b** Jam queue propogation. **c** Jam queue collision

6.4.3 The Analytical Solution for Phase Diagram

In this subsection, we derive the analytical boundaries of phases for the scenario of an on-ramp merging bottleneck based on the proposed spatial-temporal queueing model.

Figure 6.19 shows three kinds of jam queues that can be generated via the above new queueing model, due to periodical ramping vehicle disturbances:

(1) When $\tau_{in} > \tau_{out}$, the jam queue width decreases. If the jam queue dissipates before the next ramping vehicle inserts into the main road, we may observe a localized jam queue; see Fig. 6.19a.
(2) When $\tau_{in} = \tau_{out}$, we can observe a moving localized jam queue; see Fig. 6.19b.
(3) When $\tau_{in} < \tau_{out}$, two jam queues finally merge into one. The jam queue width increases, see Fig. 6.19c.

According to Helbing et al. (2009), Schönhof and Helbing (2007), Treiber et al. (2000, 2010), there are six possible congestion patterns in this scenario: Free Traffic (FT), Pinned Localized Cluster (PLC), Moving Localized Cluster (MLC), Stop-and-Go Waves (SGW), Oscillatory Congested Traffic (OCT), and Homogeneous Congested Traffic (HCT). In this subsection, we derive the analytical emerging conditions for these patterns based on the above queueing model.

Noticing that the proposed model gives prominence to three independent parameters: τ_{in}, τ_{out} and T, if assume τ_{out} be a constant as suggested in Mahnke and Kaupužs (2001), Mahnke et al. (2005), Mahnke and Kühne (2007), Mahnke and Pieret (1997), we can plot the congestion phase diagram in the widely adopted q_{ramp} versus q_{main} plot, due to the relations between τ_{in} and q_{main}, T and q_{ramp}.

First, as proven in Chiabaut et al. (2009, 2010), when a vehicle from upstream meet the first jam queue, the join time τ_{in} with a given inflow rate q_{main} should be

$$\tau_{in} = \frac{v_{free}}{q_{main}(v_{free} + w)} = \frac{v_{free} h_{main}}{v_{free} + w} \tag{6.71}$$

where $h_{main} = 1/q_{main}$ is the main road headway.

Second, q_{ramp} is the reciprocal of T by definition

$$q_{ramp} = 1/T \tag{6.72}$$

Since ramping become impossible when T is too small, we will only discuss the scenario of $T \geqslant 2\tau_{out}$ in the rest of this section. In the T versus τ_{in} phase diagram, we analytically classify the traffic phases into the following four main categories:

6.4.3.1 Case (1) FF

We define the free-flow state as that a jam queue dissipates within a very short time period. Clearly, only if $\tau_{in} \gg \tau_{out}$ and T is large enough, we get the free-flow traffic state.

Figure 6.20a shows the case in which the ramping vehicle affects only one main road vehicle. The critical inflow headway in the free-flow case can be defined as

$$h_{free} = 2\tau_{out} - \frac{s_{safe}}{v_{free}} = \frac{2v_{free} - v_{cong}}{v_{free}} \cdot \tau_{out} \tag{6.73}$$

where h_{free} guarantees the mainline following vehicle being not influenced even in the worst case of a sudden deceleration of the leading mainline vehicle.

Thus, as shown in Fig. 6.20a, the phase condition for such a free-flow scenario is

$$\tau_{in} \geqslant v_{free} h_{free}/(v_{free} + w) \tag{6.74}$$

6.4.3.2 Case (2) FF, PLC and SGW

If the dissipation time of a jam queue becomes larger, more vehicles will be disturbed by more tha one jam queue; see Fig. 6.19a. According to the asymptomatic property of $\tau_{in}^{(k)}$ given by Eq. (6.65), for those vehicles that encounter more tha one jam queue, their join times decrease as $\tau_{in}^{(k+1)} < \tau_{in}^{(k)}$, with $\tau_{in}^{(1)} = \tau_{in}$. This might make the dissipation time of the second jam queue larger than that of the first jam queue. If the increase of the dissipation time converges to a certain limit, we may observe a localized congestion. Otherwise, we will observe a sequence of jam queues whose influenced road regions extend spatially from one to another.

According to Fig. 6.19a, if the first jam queue dissipates after just p vehicles, $p = 2, \ldots, n$, we have

$$\begin{cases} \tau_{\text{in}} > \tau_{\text{out}} \\ (p-1)\tau_{\text{in}} < p\tau_{\text{out}} \\ p\tau_{\text{in}} \geqslant (p+1)\tau_{\text{out}} \end{cases} \qquad (6.75)$$

Then

$$\frac{\tau_{\text{out}}}{\tau_{\text{in}} - \tau_{\text{out}}} \leqslant p < \frac{\tau_{\text{in}}}{\tau_{\text{in}} - \tau_{\text{out}}} \qquad (6.76)$$

As shown in Fig. 6.20c, we further assume there are r vehicles influenced by only one or none jam queues when passing the ramping region before the second ramping vehicle inserting into the mainline, we have

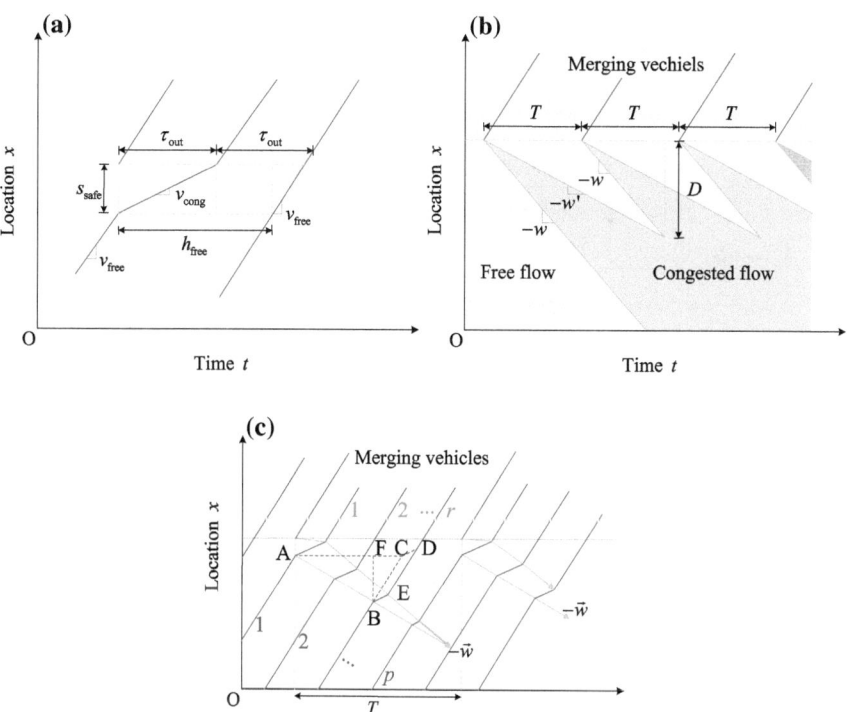

Fig. 6.20 Critical conditions of free flow and congested flow. **a** One special scenario of free flow. **b** Appropriate colliding points. **c** One special scenario of PLC/SGW

$$T \geqslant \frac{(r-1)w\tau_{in}}{v_{free}} + r\tau_{out} + h_{min} \tag{6.77}$$

where the first term on the right side of Ineq.(6.77) corresponds to $|CF| = t_C - t_F$, t_C and t_F mean the time coordinates of points C and F, respectively; the second term represents $t_E - t_A$, so the first two terms denote $t_D - t_A$ that is approximately the arrival time of the rth vehicle at the on-ramp location; the third term of h_{min} guarantees the minimal headway between the merging vehicle and the rth mainline vehicle.

Thus, Ineq.(6.77) formulates the boundary condition that at least r vehicles pass the merging region without influences by the second merging vehicle, or equivalently

$$r \leqslant \frac{w\tau_{in} + v_{free}(T - h_{min})}{w\tau_{in} + v_{free}\tau_{out}} \tag{6.78}$$

Define the proportion of vehicles that are only influenced by one jam queue as

$$\varphi \equiv \frac{r}{p} \tag{6.79}$$

Substitute Eq.(6.79) into Ineq.(6.76) and Ineq.(6.78), and eliminate the intermediate variable r and p, we obtain the following condition by neglecting the relatively small influence of h_{min}

$$T \geqslant \frac{-w\tau_{in}^2 + (\varphi+1)w\tau_{out}\tau_{in} + \varphi v_{free}\tau_{out}^2}{v_{free}(\tau_{in} - \tau_{out})} = f(\varphi, \tau_{in}) \tag{6.80}$$

where $f(\varphi, \tau_{in})$ is the boundary function with respective to φ and τ_{in}.

Clearly, when $\varphi = 1$, we get the FF phase with the boundary

$$T = \frac{-w\tau_{in}^2 + 2w\tau_{out}\tau_{in} + v_{free}\tau_{out}^2}{v_{free}(\tau_{in} - \tau_{out})} = -\frac{w}{v_{free}}(\tau_{in} - \tau_{out}) + \frac{(v_{free} + w)\tau_{out}^2}{v_{free}(\tau_{in} - \tau_{out})} \tag{6.81}$$

When $0 < \varphi < 1$, the rest $p - r$ vehicles that depart from the first jam will join the second jam queue with join time $\tau_{in}^{(2)}$, while the other vehicles join the second jam with join time $\tau_{in}^{(1)}$. We can then get the number of the vehicles influenced by the second jam queue as

$$\frac{\tau_{out} + (p-r)(\tau_{in} - \tau_{in}^{(2)})}{\tau_{in} - \tau_{out}} \leqslant q < \frac{\tau_{in} + (p-r)(\tau_{in} - \tau_{in}^{(2)})}{\tau_{in} - \tau_{out}} \tag{6.82}$$

where the second jam queue dissipates after just q vehicles.

Based on Eq.(6.67), we could have

$$\frac{\tau_{out}}{\tau_{in} - \tau_{out}} + (p-r)(1-a) \leqslant q < \frac{\tau_{in}}{\tau_{in} - \tau_{out}} + (p-r)(1-a) \tag{6.83}$$

which indicates that if $(p - r)(1 - a) \geq 1$, we will have $q > p$. Similar deductions show that when $0 < \varphi < 1$, the number of the vehicles influenced by the $(k + 1)$th jam queue will always be larger than the number of vehicles influenced by the kth jam queue. Further tests show that when $\varphi < 0.8$, the increase of the dissipation time will have no limits. For simplicity, we choose a threshold φ_{PLC} in Ineq. (6.77) to distinguish PLC and SGW phases, satisfying

$$0 < \varphi_{PLC} < \varphi_{FF} = 1 \tag{6.84}$$

6.4.3.3 Case (3) MLC

In this case, we define MLC as a single moving jam queue that propagates upstream with a constant size. Based on the above discussion, we can easily find one condition for τ_{in} as

$$\tau_{in} = \frac{v_{free}}{q_{main}(v_{free} + w)} = \tau_{out} \tag{6.85}$$

This scenario is quite a special one, because τ_{in} keeps the constant and the same as τ_{out}. But we have to set a relatively large $T \gg \tau_{out}$ to guarantee that the distance between two consecutive moving clusters is large enough; otherwise the obtained congestion pattern might be recognized as oscillating jams.

6.4.3.4 Case (4) SGW, OCT, and HCT

The phases of SGW/OCT/HCT occur when $\tau_{in} < \tau_{out}$; see Fig. 6.20b. More preciously, according to the convective instability and road capacity constraint, we have

$$\tau_{in} \geq \max \left\{ \frac{v_{cong}\tau_{out}}{v_{cong} + w}, \frac{v_{free}}{q_{max}(v_{free} + w)} \right\} \tag{6.86}$$

In order to study their dividing boundaries, we would like to check the propagation difference of acceleration wave. We have

$$w' = \underbrace{(x_{a,i-1}^{k} - x_{a,i}^{k})/\tau_{out}}_{\forall k, \, i} = (v_{cong} + w)\frac{\tau_{in}}{\tau_{out}} - v_{cong} \geq 0 \tag{6.87}$$

where w' is the acceleration wave speed; see Fig. 6.20b.

The propagation distance for the downstream front of the previous jam queue before it collides with the upstream front of the next jam queue can be estimated as

$$D \approx \frac{wT[(v_{cong} + w)\tau_{in} - v_{cong}\tau_{out}]}{(v_{cong} + w)(\tau_{out} - \tau_{in})} > 0 \tag{6.88}$$

If D is relatively large, the small separate jam queues spend more time to merge into a larger jam queue. And we can see a SGW phase. If D becomes smaller, then the OCT phase appears; if D becomes even smaller, the HCT phase emerges.

Let us define the following dimensionless scaling parameter to compare D with $w\tau_{out}$ as

$$\theta = \frac{D}{w\tau_{out}} \tag{6.89}$$

From Eq. (6.88), we have the following monotonously decreasing relationship between τ_{out} and T

$$\tau_{in} = \left(1 - \frac{wT}{(v_{cong}+w)(\theta\tau_{out}+T)}\right)\tau_{out} = g(\theta, T)\tau_{out} \tag{6.90}$$

where $g(\theta, T)$ is the substitution function with respective to θ and T.

Clearly, we have

$$\lim_{T\to 0^+} \tau_{in} = \tau_{out}, \quad \lim_{T\to\infty} \tau_{in} = \frac{v_{cong}\tau_{out}}{v_{cong}+w} \tag{6.91}$$

Choose two thresholds θ_{OCT} and θ_{HCT} in Eq. (6.90) to distinguish OCT and HCT phases, satisfying

$$0 < \theta_{HCT} < \theta_{OCT} \tag{6.92}$$

We predict that when $g(\theta_{HCT}, T)\tau_{out} < \tau_{in} \leqslant g(\theta_{OCT}, T)\tau_{out}$, we get the OCT phase; when $\tau_{in} \leqslant g(\theta_{HCT}, T)\tau_{out}$, we get the HCT phase instead; otherwise, we get the SGW phase.

In summary, Fig. 6.21a shows the congested traffic patterns denoted by the merging time interval T and the join time τ_{in}. Furthermore, we transfer the T versus τ_{in} phase diagram into the regular q_{ramp} versus q_{main} phase diagram, see Fig. 6.21a. As illustrated in Fig. 6.21, we formulate the six regions in the analytical phase diagram: (1) FF; (2) PLC; (3) MLC; (4) SGW; (5) OCT; (6) HCT. The dashed curves indicate the boundaries of each region. We adopt two dimensionless scaling parameters φ and θ to distinguish the phases of FF/PLC/SGW and SGW/OCT/HCT, respectively.

It should be pointed out that we have two kinds of SGW patterns here. In the scenarios with the condition of $\tau_{in} > \tau_{out}$, the jam queues finally dissipate but their lasting time periods will become larger one by one. In the scenarios of $\tau_{in} < \tau_{out}$, the jam queues finally merge into a wide jam that will last forever.

The above boundaries are not accurate, since we only discuss the evolution dynamics of the first few queues and meanwhile employ several approximations. So in the next section, we use simulations to verify the accuracy of the predicted phase transition boundaries.

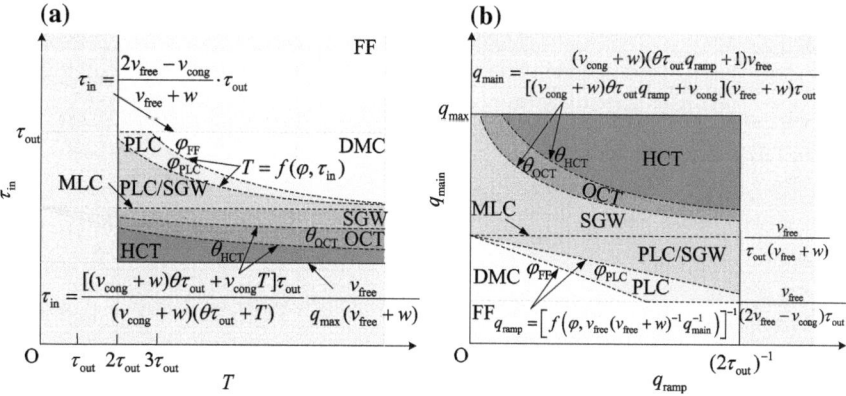

Fig. 6.21 Analytical phase diagram of the spatial-temporal queueing model. **a** T versus τ_{in}. **b** q_{ramp} versus q_{main}

6.4.4 Numerical Example

6.4.4.1 Parameter Setting for the Ramping Scenario

To verify the above analytical derivation of phase boundaries, we discuss a typical ramping scenario that had been addressed in literature (Chen et al. 2010b; Lu et al. 2010; Treiber et al. 2010). The studied ramping region is with a bottleneck at $x = 7.5$ km. Compared to the ramping region studied in Lu et al. (2010), the geometric feature of the ramping road is neglected.

We assume that all the vehicles in the simulations are small cars. As suggested in the literature (Chen et al. 2010b; Helbing et al. 2009; Kerner 2009, 2004; Kim and Zhang 2008; Lu et al. 2010; Schönhof and Helbing 2007; Treiber et al. 2000, 2010). We set $v_{free} = 90$ km/h $= 25$ m/s, $w \approx 18$ km/h $= 5$ m/s, and $\tau_{out} = 2$ s as suggested in Mahnke and Kaupužs (2001), Mahnke et al. (2005), Mahnke and Kühne (2007), Mahnke and Pieret (1997). Table 6.3 lists other parameters that are used in the simulations.

We can hold the convective condition, because $q_{main} \leqslant q_{max} = 2/3$ veh/s, $\tau_{in} \geqslant v_{free} q_{max}^{-1}(v_{free} + w)^{-1} = 1.25$ s and $v_{cong} \tau_{out}(w + v_{cong})^{-1} = 0.95$ s. Moreover, we can also guarantee the collision free condition $s_{cong}^{(1)} > s_{safe}$, because $\tau_{in}^{(1)} > v_{cong} \tau_{out}/(v_{cong} + w) = 1.05$ s.

6.4.4.2 On-Ramp Merging Rule

We assume that the mainline vehicles have higher priority than the merging vehicles. If possible, one merging vehicle will be inserted into the mainline, strictly according to the time period T. However, if doing this leads to an immediate collision between

Table 6.3 Specific parameters for numerical example

Parameters	Symbols	Values
Main road capacity	q_{max}	2,400 veh/h/lane
Main road free flow speed	v_{free}	90 km/h
Main road congested flow speed	v_{cong}	20 km/h
Congestion wave propagation speed	w	18 km/h
Maximal traffic flow spacing	s_{max}	50 m
Minimal unaffected spacing	s_{min}	37.5 m
Safety spacing	s_{safe}	11.1 m
Departing time	τ_{out}	2 s
Dimensionless scaling coefficient for FF	φ_{FF}	1.0
Dimensionless scaling coefficient for PLC	φ_{PLC}	0.8
Dimensionless scaling coefficient for OCT	θ_{OCT}	10
Dimensionless scaling coefficient for HCT	θ_{HCT}	5

the merging vehicle and the mainline vehicle that just leaves the ramping point, we will let the merging vehicle wait for a while and then merge into the mainline. We call it intelligent on-ramp merging rule or minimal headway rule.

In other words, the merging vehicle will wait until its headway to the leading vehicle is larger than the minimal critical headway $h_{min} = 1/q_{max}$. This assumption is consistent with our driving experiences and has been adopted by many previous simulation systems (Hidas 2005). The intelligent merging behavior will prevent generating unsafe spacings in simulations, but it will certainly result in more complicated merging disturbances and thus more complex congestions. It will finally influence the simulated phase diagram, especially for FF, PLC and MLC phases.

It should be pointed out that no lane changing actions are considered, mainly because the model allows only two kinds of vehicles speeds. So, the vehicle will not benefit from lane changing actions. It is worthy to point out that the spatial-temporal queueing model is highly suitable for a simulation approach because we only consider the transition points and ignore the dynamics between two adjacent points. This technique enables us to reproduce the spatial-temporal evolution of traffic speed much more quickly.

6.4.4.3 Numerical Results

Figure 6.22 shows the typical traffic patterns when we choose different values of q_{ramp} and q_{main}. In order to get a uniform grid in the widely adopted q_{ramp} versus q_{main} phase diagram, we sample the parameter space of (q_{ramp}, q_{main}) uniformly, with a discretization level of 50 veh/h for q_{ramp} and another discretization level of 100 veh/h for q_{main}, respectively. Thus, the sampling grid in the T versus τ_{in} phase diagram is not uniform due to the reciprocal relationships.

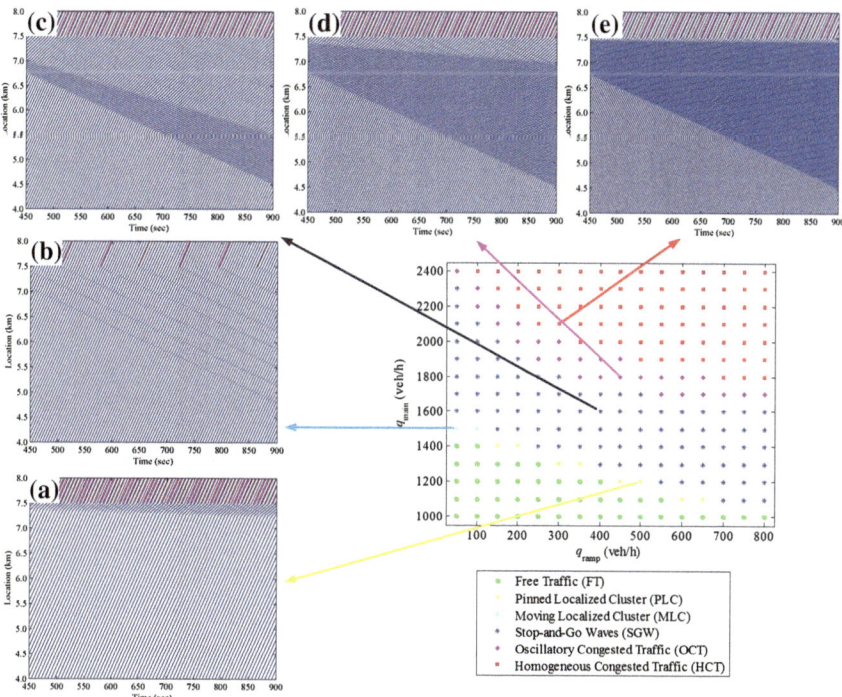

Fig. 6.22 The obtained phase diagram. 5 states: Pinned Localized Cluster (PLC), Moving Localized Cluster (MLC), Stop-and-Go Wave (SGW), Oscillatory Congested Traffic (OCT), and Homogeneous Congested Traffic (HCT), are reproduced. The selected conditions are: **a** $(q_{ramp}, q_{main}) =$ (500, 1200) veh/h. **b** $(q_{ramp}, q_{main}) =$ (50, 1500) veh/h. **c** $(q_{ramp}, q_{main}) =$ (400, 1600) veh/h. **d** $(q_{ramp}, q_{main}) =$ (450, 1800) veh/h. **e** $(q_{ramp}, q_{main}) =$ (300, 2100) veh/h

Figure 6.22 also presents the corresponding trajectory plots for five representative congestion states. In order to make the plots more clear, we only draw the trajectory of the first vehicle for every two consecutive vehicles; otherwise the curves will be too dense to observe. The red lines, which start at the location $x = 7.5$ km in the vehicle trajectory plots, denote the merging on-ramp vehicles. The headway between two adjacent red lines represents the merging time interval.

Figure 6.23 compares the predicated boundaries obtained in the analytical and the simulated boundaries. Here, the dimensionless scaling parameters are chosen as $\varphi_{FF} = 1$, $\varphi_{PLC} = 0.8$, $\theta_{OCT} = 10$ and $\theta_{OCT} = 5$. Clearly, the boundaries of FF, PLC, MLC, SGW, OCT, and HCT phases fit dramatically well with those estimated in analytical results.

The MLC states only appear at $(q_{ramp}, q_{main}) = (50, 1500)$ veh/h and $(q_{ramp}, q_{main}) = (100, 1500)$ veh/h that correspond to $T = 72$ s and $T = 36$ s. Because the maximum T is 18 s in Fig. 6.23a, we cannot see the MLC pattern in this subfigure.

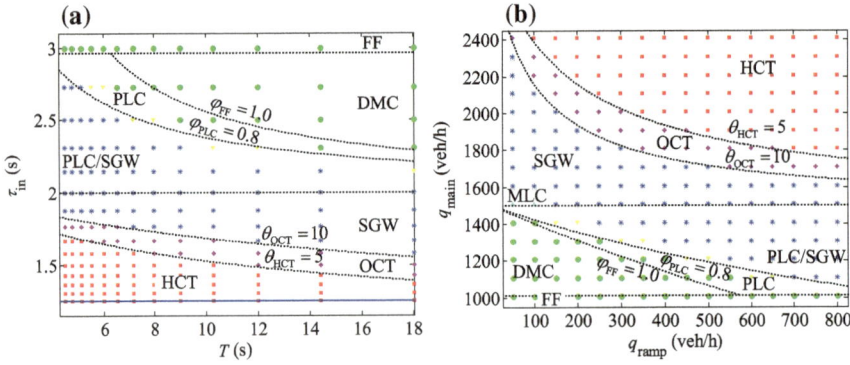

Fig. 6.23 Comparison between theoretical and simulated phase diagram. **a** T versus τ_{in}. **b** q_{ramp} versus q_{main}

The locations of congestion patterns are similar to those that have been obtained via different simulation models. This shows the effectiveness of such a simple spatial-temporal queueing model.

As revealed in the phase condition, any simulation model that can reproduce PLC patterns should be able to generate a roughly equal jam queue join time τ_{in} and departing time τ_{out}. Thus, we can directly analyze the mainline-ramping flow rate condition that allows the $\tau_{in} \approx \tau_{out}$.

It should be pointed out that the SGW (with $\tau_{in} < \tau_{out}$) and OCT congestion patterns that are generated do not strictly reproduce the SGW and OCT congestion patterns that have been observed in practice. Particularly, in our model, the narrow jams in which vehicles move with a low speed cannot merge in to a large jam where vehicles fully stop. This is mainly because our model allows only two kinds of vehicles speeds for simplicity. Indeed, the SGW/OCT trajectories shown in Fig. 6.23 are similar to the trajectories obtained via the OVM in Wagner and Nagel (2008). This indicates that the equivalence between Newell's simplified model and the OVM in terms of phase diagram.

In this section, we propose an as simple as possible spatial-temporal queueing model to study the phase diagram around ramping regions. The merits of this new model include but are not limited to:

(1) This model adopts a minimum set of parameters and rules to analytically formulate the traffic flow phase diagram. All the parameters have clear physical meanings. Therefore, it is easy to analyze. Different from previous approaches, this model could derive analytical boundaries for the phases. One major concern of this approach is to find what a phase diagram can be gotten if only a simple density-speed relation model is adopted. Surprisingly, FF/PLC/MLC/HCT patterns can be well reproduced even via such a simple model. However, results also reveal that the occurrence and evolution of SGW/OCT patterns still needs further discussions.

(2) Due to its conciseness, this model reveals the basic factors that govern the instable nature of traffic flow in on-ramp regions. On the microscopic level, these factors are: the time τ_{in} for a vehicle to join the jam queue and the merging time interval T. Noticing the tight relationship between τ_{in} and the headway/spacing of a vehicle, we can see that the dominating factors of macroscopic traffic congestion patterns are the microscopic headways/spacings adopted by drivers with respect to mainline inflow rate. Indeed, the proposed model gives a simplified yet equivalent description of the close tracking behaviors that were depicted via many other stimulus-response type or optimal-velocity type car-following models. But the conciseness of the proposed method, which inherits from Newell's simplified car-following model, makes it distinct from other approaches. On the macroscopic level, the corresponding factors are the main road flow rate q_{main} and ramping road flow rate q_{ramp}. Based on this queueing model, we show how τ_{in} and q_{main}, as well as T and q_{ramp}, depend on each other. As a result, we link microscopic individual vehicle movements with macroscopic congestion patterns.

However, this new model is not perfect. Particularly in the following aspects:

(1) This model does not consider the stochastic features of τ_{in} and τ_{out} in reality. Thus, some stochastic properties of traffic flow cannot be reproduced via this model.
(2) This model allows only two kinds of vehicles speeds and forbids lane change for simplicity. Although this simplification helps emphasize the major driven force of congestion evolution, it meanwhile prevents the appearances of nonhomogeneous jam queues. And the obtained SGW and OCT phases are not exactly the same as the ones reproduced by the Gas-Kinetic-based Traffic model (GKT model) and Intelligent Driver Model (IDM). We plan to modify the model to solve this problem in our coming reports.
(3) This model implicitly adopts the hypothesis that the speed and density satisfy a linear relationship in the congested flows (Daganzo 2005; Lighthill and Whitham 1955; Richards 1956). Therefore, it cannot explain some nonlinear effects in congestion formation. How to overcome this shortcoming needs further discussions.

In short, this new model provides a reasonable starting point for modeling more complex traffic phenomena.

6.5 Discussions

Traffic flow breakdown phenomena occur with continuous oscillations and lead to a wide range of spatial-temporal traffic congestions. Based on the analysis of the stochastic traffic capacity, this chapter focuses on the transition process from perturbations to traffic jams in the metastable traffic flow, and studies the varying features of TF-BP under the framework of vehicle trajectory analysis. Unlike previous studies, this chapter discusses the close relationship between TF-BP and the jam wave propagation phenomena with concentration on the evolution of jam queue size.

Furthermore, the simplified G/D/1 queuing model and general G/G/1 queuing model are proposed based on Newell's simplified car-following model, and they link the microscopic driving behaviors with the dynamic characteristics of the corresponding macroscopic traffic flow patterns. The formation and dissipation of jam queues are treated as comprehensive effects of stochastic joining and leaving processes for upstream and downstream vehicles, respectively. Simulation results show that the estimated TF-BP curves well fit with the widely used Weibull distributions.

Finally, the chapter proposes a concise spatial-temporal queueing model to describe the formation mechanism and evolution process of traffic jams based on the stochastic Newell's condition in the ramping bottleneck on freeways. We analytically derive the phase diagram that is useful for presenting the evolution process for different traffic flow phases and quantitatively determining the phase transition conditions. Simulations results show two main congestion modes: *spatially localized congestion* and *spatially extended congestion*, they are helpful to reveal the inherent mechanism of phase transitions.

From the perspective of field applications, studies of TF-BP in bottleneck areas focus on the increase in traffic capacity by analyzing the nature, intensity, and evolution of traffic congestions. In the future studies, optimal control models of active traffic management (ATM) can be established in order to minimize TF-BP and maximize the expectation of stochastic traffic capacity. With a careful consideration of the ramping influences on the mainline, feasible, dynamic, and optimal control schemes (strategies) to alleviate bottleneck congestions include:

(1) *Ramp metering* Investigate the influences of different ramping signal control strategies on the mainline traffic flow, and obtain the optimal solutions of ramp metering strategies by adjusting the ramping flows conditional on a maximal acceptable value of TF-BP.

(2) *Mainline variable speed limits* Study time-varying speed limits of the upstream mainline and their influences on traffic state evolutions. Establish the optimization model of variable speed limits to enhance the mainline capacity constrained by the upper limit of TF-BP threshold.

(3) *Lane-changing control* After entering a weaving section, vehicles need to complete lane-changing maneuvers that result in stochastic movements and reduce the expected capacity of weaving sections. Study the entering/exiting behaviors on weaving sections and their impacts on TF-BP and stochastic capacity in order to reduce lane-changing-related TF-BP and improve safety and level of service.

Chapter 7
Conclusions and Future Work

The main efforts and contributions of this book are briefly outlined as follows:

- **Summaries of stochastic traffic flow modeling on the basis of the historical development** In Chap. 2, we reviewed and summarized the history of traffic flow theory in terms of macroscopic, mesoscopic, and microscopic modeling approaches. Four categories of stochastic approaches were summarized for continuous traffic flow modeling: randomization of first order conservation law and higher-order momentum equations; randomization of driving behaviors and mixed traffic flow; randomization of relationships among flow, density and velocity; randomization of road capacity and travel time distribution in transportation reliability studies. Particularly, we focused on the probabilistic distributions of headway/spacing/velocity, based on which comprehensive methods such as Markov models, stochastic fundamental diagram, stochastic Newell condition would be proposed to better fit the empirical observations.

- **Empirical observations of complex, dynamic, and stochastic evolutions of traffic flow on the basis of two dominating measurements** In Chap. 3, we made use of two typical kinds of datasets (*Eulerian* and *Lagrangian* field measurements) and several approaches (joint/marginal/conditional probabilities, statistics, signal processing) for the qualitative and quantitative analysis of traffic flow evolution characteristics to reveal the complex, dynamic and stochastic phenomena. Specifically, fixed loop detectors data were used as *Eulerian* measurements and NSGIM trajectory data were used as *Lagrangian* measurements. Three significant aspects of stochastic traffic flow features were studied, including headway/spacing/velocity distributions, disturbances of congested platoons (jam queues) and time-frequency properties of oscillations. The observations deepen our understanding of traffic flow characteristics.

- **Development of the headway/spacing Markov model and asymmetric stochastic Tau theory in car-following behaviors** In Chap. 4, we linked mesoscopic headway/spacing distributions with a unified car-following model based on Markov state transition theory. The parameters of this model were directly estimated from NGSIM data. The empirical headway/spacing distributions were viewed as

© Tsinghua University Press, Beijing and Springer-Verlag Berlin Heidelberg 2015   163
X. (M.) Chen et al., *Stochastic Evolutions of Dynamic Traffic Flow*,
DOI 10.1007/978-3-662-44572-3_7

outcomes of stochastic car-following behaviors and reflections of unconscious and inaccurate perceptions people had. To explain the phenomenon that the observed headways follow a certain lognormal type distribution in terms of velocity, an asymmetric stochastic extension of the well-known Tau Theory was proposed. The agreement between the model predictions and the empirical observations indicated that the physiological Tau characteristics of human drivers govern driving behaviors in an implicit way.

- **Examination of wide scattering characteristics in stochastic fundamental diagram** In Chap. 5, we discussed the implicit but tight connection between headway/spacing distributions and the wide scattering feature of flow-density plot based on the Newell's simplified car-following model. We first examined microscopic driving behaviors that were retrieved from the NGSIM trajectory database. Results showed that asymmetric driving behaviors result in a family of velocity-dependent lognormal type spacing/headway distributions. Based on this fact, we then proposed a probabilistic model to characterize the 2D region of congested flows in spacing-velocity plot. Extending the Newell's simplified car-following model, we finally discussed the corresponding 2D region of congested flows in the flow-density plot. Results surprisingly revealed that the seemingly disorderly scattering points on the flow-density plot had a certain distribution tendency. Indeed, such tendency was determined by the microscopic headway/spacing distributions.

- **Traffic flow breakdown probability estimation in highway bottlenecks: queueing theory and phase diagram analysis** In Chap. 6, traditional approach of non-parametric statistics to the TF-BP problem was well developed for the usage of *Eulerian* traffic flow measurements. In order to explore the emerging *Lagrangian* data, a simple G/D/1 queueing model and a more general G/G/1 queueing model were proposed on the basis of Newell's simplified car-following model. We showed that the growth and dissipation of a jam queue were controlled by two competing stochastic processes: the upstream vehicle joining process and downstream vehicle departing process. Monte-Carlo simulations were incorporated to estimate TF-BP. Queueing model could be further used to help design ramp metering strategy and cooperative driving strategy to reduce/avoid traffic breakdowns. We also derived the analytical phase diagram for the evolution process of different traffic flow phases and determined phase transition conditions, which helped reveal that the dominating factors in the formations and transitions of different congestion patterns were: the time τ_{in} for a vehicle to join the jam queue and the merging time interval T (microscopic level); main road flow rate q_{main} and ramping road flow rate q_{ramp} (macroscopic level).

Several other significant explorations have not been included in this work, which may be the focus of future research efforts.

- **Data fusion of multisource and heterogeneous sensors** The *Eulerian* measurements consist of traffic flow counts, speed and occupancies at fixed locations, such as loop detectors, locally virtual video cameras, AVI and LPR records; while the *Lagrangian* measurements identify specific vehicle trajectories by using large-scale video cameras or radars, GPS equipped vehicle, GPS equipped mobile devices (location based technology inside cellular phones), tracking device, etc. As the data availability of both measurements continue to rapidly expand, it may be possible to engage on the data fusion of multisource and heterogeneous sensors, and derive more accurate traffic flow models and algorithms for stochastic characteristics analysis that may be beneficial in traffic state estimation and prediction. Since the measured quantities, the spatial and temporal extent of the two kinds of measurements are generally various but complementary to each other, data fusion of heterogeneous sensors would be useful in areas without dedicated sensors deployed.

- **Development and implementation of TF-BP incorporated control strategies**
  In field applications of the TF-BP estimation approach and highway bottleneck phase diagram analysis, the capacity is regarded stochastic. In order to increase the expectation of traffic capacity in bottlenecks and minimize the probability of congestion evolutions, optimal control strategies of active traffic management (ATM) could be established to alleviate bottleneck congestions with the consideration of TF-BP constraint, including ramp metering; mainline variable speed limits; lane-changing behaviors control. This spatial-temporal queueing model implicitly adopted the hypothesis of a linear relationship of speed and density in the congested flows. How to explain more nonlinear effects in congestion formation, and how the analytical phase diagram would change when randomizing τ_{in} and τ_{out} still need more further research efforts.

Appendix A
Linear Stability Analysis of the Higher-Order Macroscopic Model

In order to analyze the linear stability properties of the general higher-order macroscopic traffic flow model of Eq. (2.14), we assume that $\rho(x, 0) = \rho_0$ and $v(x, 0) = v_0$ are the steady state solutions, respectively. Define the following variations

$$\begin{cases} \delta\rho = \delta\rho(x, t) = \rho(x, t) - \rho_0 \\ \delta v = \delta v(x, t) = v(x, t) - v_e(\rho_0) \end{cases} \tag{A.1}$$

Substitute $\rho(x, t) = \rho_0 + \delta\rho$ and $v(x, t) = v_e(\rho_0) + \delta v$ into Eq. (2.14), it can be expanded in Taylor's series, ignore the nonlinear terms, we obtain the following linear equations

$$\frac{\partial\delta\rho}{\partial t} + v_e(\rho_0)\frac{\partial\delta\rho}{\partial x} + \rho_0\frac{\partial\delta v}{\partial x} = 0 \tag{A.2}$$

$$\frac{\partial\delta v}{\partial t} + v_e(\rho_0)\frac{\partial\delta v}{\partial t} = \frac{v_e(\rho_0) - v}{\tau} + k_1 v'_e(\rho_0)\frac{\xi}{\tau}\frac{\partial\delta\rho}{\partial x} + k_2\frac{\eta}{\tau}\frac{\partial\delta v}{\partial x} \tag{A.3}$$

Assume that the perturbations on traffic flow density and speed are exponential

$$\begin{cases} \rho(x, t) = \rho_0 + \tilde{\rho}e^{ikx + \omega(k)t} \\ v(x, t) = v_0 + \tilde{v}e^{ikx + \omega(k)t} \end{cases} \tag{A.4}$$

Thus,

$$\frac{\partial\delta\rho}{\partial x} = ik\tilde{\rho}e^{ikx + \omega(k)t}, \quad \frac{\partial\delta\rho}{\partial t} = \omega\tilde{\rho}e^{ikx + \omega(k)t}$$

$$\frac{\partial\delta v}{\partial x} = ik\tilde{v}e^{ikx + \omega(k)t}, \quad \frac{\partial\delta v}{\partial t} = \omega\tilde{v}e^{ikx + \omega(k)t}$$

© Tsinghua University Press, Beijing and Springer-Verlag Berlin Heidelberg 2015
X. (M.) Chen et al., *Stochastic Evolutions of Dynamic Traffic Flow*,
DOI 10.1007/978-3-662-44572-3

Substitute Eq. (A.4) into Eq. (2.14), we have

$$\mathbf{A}\tilde{X} = 0 \tag{A.5}$$

where $\tilde{X} = (\tilde{\rho}, \tilde{v})^T$, the elements of matrix $\mathbf{A}$ are

$$A_{11} = \omega + ikv_0$$
$$A_{12} = ik\rho_0$$
$$A_{21} = \frac{v'_e(\rho_0)}{\tau} + ikk_1\frac{\xi}{\tau}v'_e(\rho_0)$$
$$A_{22} = -(\omega + ikv_0) - \frac{1}{\tau} + ikk_2\frac{\eta}{\tau}$$

Linear equations in Eq. (A.5) have valid solutions only if $\det \mathbf{A} = A_{11}A_{22} - A_{12}A_{21} = 0$, then we get a quadratic plural equation with respect to ω as follows

$$\omega^2 + \left(\frac{1}{\tau} + i\left(2kv_0 - kk_2\frac{\eta}{\tau}\right)\right)\omega$$
$$+ \left(k^2 v_0 k_2 \frac{\eta}{\tau} - k^2 v_0^2 - k^2 \rho_0 v'_e(\rho_0) k_1 \frac{\xi}{\tau}\right)$$
$$+ i\frac{k\left(v_0 + \rho_0 v'_e(\rho_0)\right)}{\tau} = 0 \tag{A.6}$$

It can be reduced to

$$\omega^2 + (\phi_1 + i\varphi_1)\omega + (\phi_2 + i\varphi_2) = 0 \tag{A.7}$$

where

$$\phi_1 = \frac{1}{\tau}$$
$$\varphi_1 = 2kv_0 - kk_2\frac{\eta}{\tau}$$
$$\phi_2 = -k^2 v_0^2 - k^2 \rho_0 v'_e(\rho_0) k_1 \frac{\xi}{\tau} + k^2 v_0 k_2 \frac{\eta}{\tau}$$
$$\varphi_2 = -\frac{k\left(v_0 + \rho_0 v'_e(\rho_0)\right)}{\tau}$$

The solutions of Eq. (A.7) are

$$\omega_\pm = \frac{-\phi_1 - i\varphi_1 \pm \sqrt{\left(\phi_1^2 - \varphi_1^2 - 4\phi_2\right) + i\left(2\phi_1\varphi_1 - 4\varphi_2\right)}}{2}$$

$$= -\frac{\phi_1}{2} - i\frac{\varphi_1}{2} \pm \sqrt{\left(\frac{\phi_1^2 - \varphi_1^2 - 4\phi_2}{4}\right) + i\left(\frac{\phi_1\varphi_1 - 2\varphi_2}{2}\right)}$$

$$= -\frac{\phi_1}{2} - i\frac{\varphi_1}{2} \pm \sqrt{\Re \pm i\,|\Im|}$$

$$= \left(-\frac{\phi_1}{2} \pm \sqrt{\frac{\sqrt{\Re^2 + \Im^2} + \Re}{2}}\right) + i\left(-\frac{\varphi_1}{2} \pm \sqrt{\frac{\sqrt{\Re^2 + \Im^2} - \Re}{2}}\right) \quad \text{(A.8)}$$

where

$$\Re = \frac{\phi_1^2 - \varphi_1^2 - 4\phi_2}{4}, \quad |\Im| = \frac{\phi_1\varphi_1 - 2\varphi_2}{2} \quad \text{(A.9)}$$

The real part of the solution in Eq. (A.8) is

$$\mathrm{Re}\,(\omega_\pm) = -\frac{\phi_1}{2} \pm \sqrt{\frac{1}{2}(\sqrt{\Re^2 + \Im^2} + \Re)} \quad \text{(A.10)}$$

Let $\mathrm{Re}\,(\omega_\pm) = 0$, which is equivalent to

$$\phi_1^2 \Re + \Im^2 = \frac{\phi_1^4}{4} \quad \text{(A.11)}$$

thus,

$$\phi_1^2 \phi_2 - \varphi_2^2 + \phi_1 \varphi_1 \varphi_2 = 0 \quad \text{(A.12)}$$

Substitute the parameters of the general higher-order macroscopic model into Eq. (A.12), we obtain the critical linear stability condition as following

$$\rho_0 v_e'\,(\rho_0) + k_1\frac{\xi}{\tau} + k_2\frac{\eta}{\tau} = 0 \quad \text{(A.13)}$$

Thus, the linear stability condition is $\mathrm{Re}\,(\omega_\pm) \leqslant 0$, i.e.

$$\rho_0 v_e'\,(\rho_0) + k_1\frac{\xi}{\tau} + k_2\frac{\eta}{\tau} \geqslant 0 \quad \text{(A.14)}$$

Substitute $\partial_\rho P_1$ and $\partial_v P_2$ into Eq. (A.14), assume they satisfy $\partial_v P_2\,(\rho, v) \leqslant 0$, $v_e'\,(\rho) \leqslant 0$ and $\eta = -\frac{\tau}{\rho}\partial_v P_2(\rho, v) \geqslant 0$. Then the linear stability constion is equivalent to the following inequation in Helbing and Johansson (2009)

$$\rho_0 \left|v_e'\,(\rho_0)\right| \leqslant \sqrt{\frac{\partial P_1}{\partial \rho} + \frac{1}{4\rho^2}\left(\frac{\partial P_2}{\partial v}\right)^2} + \frac{1}{2\rho}\left|\frac{\partial P_2}{\partial v}\right| \quad \text{(A.15)}$$

Appendix B
Linear Stability Analysis of the Multi-Anticipative Car-Following Models

The purpose of microscopic traffic flow stability analysis is to study the perturbation effects from one vehicle to another vehicle. The equilibrium state of traffic flow is assumed when a platoon moves at the same spacing and speed. A general conclusion of the multi-anticipative car-following model is that the stability region increases with the decline of drivers reaction delays and the multi-vehicle information. Information of multi-vehicle speed differences and spacings increase stability in the evolution of traffic flow. Stability can be categorized into local stability and asymptotic stability. The former type accounts for car-following behaviours of two consecutive vehicles, while the latter one is to study the complex behaviours of a platoon.

When a perturbation occurs to the first vehicle or in the middle of a platoon, it may be enlarged or reduced when propagating upstream. If traffic flow is unstable, the perturbation will gradually change the smooth flow into a congested state; If the system is stable, the perturbation will gradually shrink and disappear, be ultimately controlled in a certain small range in the propagation process, after that, the system will restore equilibrium. Therefore, traffic congestion can be regarded as a phenomenon of instability, the related studies are of great significance in the analysis of traffic flow characteristics, reasons of rear-end accidents and traffic jams, increase of driving safety and road capacity etc.

Although there are a variety of car-following models in literature, most of them can be basically assorted as a nonlinear relationship between acceleration of the nth vehicle and speed, spacing and speed difference. A general expression is

$$a_n = f(x_n, ..., x_{n-m+1}; v_n, ..., v_{n-m+1}), \ m \in \mathbb{N}^+ \tag{B.1}$$

Assume a platoon runs on a homogeneous road, given an initial equilibrium state that the spacing is a constant of $\bar{s}$, the corresponding speed of each vehicle is $\bar{v}$. For N vehicles, we have

$$\bar{x}_n(t) = (N - n)\bar{s} + \bar{v}t, \ n \in \mathbb{N}^+ \tag{B.2}$$

where $\bar{s}$ and $\bar{v}$ are the net gap of two adjacent vehicles and the velocity of vehicles in homogenous flow, respectively. $\bar{x}_n(t)$ is the location of the nth vehicle at time t.

© Tsinghua University Press, Beijing and Springer-Verlag Berlin Heidelberg 2015
X. (M.) Chen et al., *Stochastic Evolutions of Dynamic Traffic Flow*,
DOI 10.1007/978-3-662-44572-3

Let $y_n(t)$ be a small perturbation with the linear Fourier-mode expanding from the steady state position of the nth vehicle at time t. The position $x_n(t)$ with as a small deviation is expressed as follows

$$x_n(t) = \bar{x}_n(t) + y_n(t) \tag{B.3}$$

Suppose the small perturbation of homogenous flow is chosen as:

$$y_n(t) = \Re\left(ce^{i\alpha_k n + zt}\right) = x_n(t) - \bar{x}_n(t), \quad |y_n(t)| \ll 1 \tag{B.4}$$

where c is the constant, $\alpha_k = 2\pi k/N$, $k = 0, ..., (N-1)$.

We use the long-wavelength expansion to analyze the three models and here we generally consider the finite reaction time. However, the instabilities can also first arise at nonzero wave number, i.e. the short-wavelength instability, for $\tau > 0$, which can be very complex. While the problem does not arise for the case of $\tau = 0$ (Wilson 2008). For simplicity, the linear analysis in the following sections only discusses the long-wavelength instability pattern.

B.1 Model I

Assume that the ac/deceleration behaviour of the nth vehicle is a linear combination of the reactions with regard to its m preceding vehicles, i.e.

$$\dot{v}_n(t + \tau) = \sum_{j=1}^{m} \gamma_j f\left(s_{n,n-j}(t), v_n(t), \Delta v_{n,n-j}(t)\right) \tag{B.5}$$

where τ is driver reaction delay. $s_{n,n-j} = x_{n-j} - x_n - L_{n-j}$ is the gap between the nth vehicle and the preceding $(n-j)$th vehicle. $\Delta v_{n,n-j}(t) = v_n - v_{n-j}$ is the velocity difference between the nth vehicle and the $(n-j)$th vehicle. L_{n-j} is the vehicle length of the $n-j$th vehicle, and equals to L for simplicity. The weighting coefficients γ_j ($j = 1, 2, ..., m$) decrease monotonically as m increases, which indicates that the influences of the vehicles ahead decrease according to the distance gaps. The weighting coefficients γ_j are introduced here, because the simply summing up formula will lead to inconsistent desired headways when considering different number of preceding vehicles at the same homogenous velocity. The inconsistence can also be avoided by selecting the suitable γ_j. We assume that $0 \leqslant \gamma_j \leqslant 1$, $\sum_{j=1}^{m} \gamma_j = 1$.

Suppose $y_n(t)$ to be a small deviation from the uniform steady state. We have

$$\ddot{y}_n(t + \tau) = \sum_{j=1}^{m} \gamma_j \left[f_{1j} \left(y_{n-j}(t) - y_{n-j+1}(t) \right) \right]$$

$$+ \sum_{j=1}^{m} \gamma_j \left[f_{2j} \dot{y}_n(t) + f_{3j} \left(\dot{y}_n(t) - \dot{y}_{n-j}(t) \right) \right] \tag{B.6}$$

where $\partial f / \partial s_{n,n-j} \, (\bar{s}, \bar{v}, 0) = f_{1j} \geqslant 0, \partial f / \partial v_n \, (\bar{s}, \bar{v}, 0) = f_{2j} \leqslant 0$ and $\partial f / \partial \Delta v_{n,n-j}$
$(\bar{s}, \bar{v}, 0) = f_{3j} \leqslant 0$.

We can rewrite Eq. (B.6) to obtain the difference equation

$$\dot{y}_n(t + 2\tau) - \dot{y}_n(t + \tau)$$

$$= \tau \sum_{j=1}^{m} \gamma_j f_{1j}(y_{n-j}(t) - y_{n-j+1}(t)) + \sum_{j=1}^{m} \gamma_j f_{2j} \left(y_n(t + \tau) - y_n(t) \right)$$

$$+ \sum_{j=1}^{m} \gamma_j f_{3j}(y_n(t + \tau) - y_n(t) - y_{n-j}(t + \tau) + y_{n-j}(t)) \tag{B.7}$$

To maintain linear stability, it requires the real part of z is negative for any mode of small perturbations $y_n(t)$ $(k = 0, 1, ..., N - 1)$. we have

$$(e^{\tau z} - 1) \left[e^{\tau z} z - \sum_{j=1}^{m} \gamma_j f_{2j} + \sum_{j=1}^{m} \gamma_j f_{3j} (e^{-ij\alpha_k} - 1) \right]$$

$$= \tau \sum_{j=1}^{m} \gamma_j f_{1j} (e^{-ij\alpha_k} - e^{-i(j-1)\alpha_k}) \tag{B.8}$$

By expanding $z = z_1(i\alpha_k) + z_2(i\alpha_k)^2 + ...$ in Taylor's series, where z is the complex growth rate, z_1 and z_2 are real numbers. Therefore, the first order and second order terms of the expression of z is given respectively as follows:

$$z_1 = \frac{\sum_{j=1}^{m} \gamma_j f_{1j}}{\sum_{j=1}^{m} \gamma_j f_{2j}} \tag{B.9}$$

$$z_2 = \frac{(1 - \tau \sum_{j=1}^{m} \gamma_j f_{2j}) z_1^2 - z_1 \sum_{j=1}^{m} \gamma_j f_{3j} j - \sum_{j=1}^{m} \gamma_j f_{1j} (j - 1/2)}{\sum_{j=1}^{m} \gamma_j f_{2j}} > 0 \tag{B.10}$$

For small disturbances with long wavelengths, the stationary traffic flow is stable when the following condition is satisfied:

$$\tau < \frac{1}{\sum_{j=1}^{m} \gamma_j f_{2j}} - \frac{\sum_{j=1}^{m} \gamma_j f_{1j} \sum_{j=1}^{m} \gamma_j f_{3j} j + \sum_{j=1}^{m} \gamma_j f_{1j} \left(j - \frac{1}{2}\right) \sum_{j=1}^{m} \gamma_j f_{2j}}{\left(\sum_{j=1}^{m} \gamma_j f_{1j}\right)^2}$$

(B.11)

It provides a criteria that can be applied into all car-following models with the consideration of reaction delays to verify the boundary of stable and unstable regimes of traffic flow.

B.2 Model II

Different from *Model I*, *Model II* takes actions according to the gap/ velocity differences between two adjacent vehicles ahead instead of the gap/velocity differences between the subject vehicle and an arbitrary preceding vehicle.

$$\dot{v}_n(t + \tau) = \sum_{j=1}^{m} \gamma_j f \left(s_{n-j+1}(t), v_n(t), \Delta v_{n-j+1}(t)\right)$$

(B.12)

where $s_{n-j+1} = x_{n-j} - x_{n-j+1} - L$ is the net gap between the $(n - j + 1)$th vehicle and the nearest preceding vehicle. $\Delta v_{n-j+1} = v_{n-j+1} - v_{n-j}$ is the velocity difference of two adjacent vehicles.

Suppose $y(t)$ to be a small disturbance from the uniform steady state, i.e.

$$\ddot{y}_n(t) = \sum_{j=1}^{m} \gamma_j [f_{1j}(y_{n-j}(t) - y_{n-j+1}(t)) + f_{2j}\dot{y}_n(t) + f_{3j}(\dot{y}_{n-j+1}(t) - \dot{y}_{n-j}(t))]$$

(B.13)

where $\gamma_1 = 1, 0 < \gamma_j < 1, j = 2, 3, ..., m$. The partial derivation of acceleration f satisfies $\partial f / \partial s_{n-j+1} (j\bar{s}, \bar{v}, 0) = f_{1j} \geqslant 0$, $\partial f / \partial v_n (j\bar{s}, \bar{v}, 0) = f_{2j} \leqslant 0$, and $\partial f / \partial \Delta v_{n-j+1} (j\bar{s}, \bar{v}, 0) = f_{3j} \leqslant 0$.

The corresponding difference equation is

$$\dot{y}_n(t + 2\tau) - \dot{y}_n(t + \tau)$$

$$= \tau \sum_{j=1}^{m} \gamma_j f_{1j}(y_{n-j}(t) - y_{n-j+1}(t)) + \sum_{j=1}^{m} \gamma_j f_{2j}(y_n(t + \tau) - y_n(t))$$

$$+ \sum_{j=1}^{m} \gamma_j f_{3j}(y_{n-j+1}(t + \tau) - y_{n-j+1}(t) - y_{n-j}(t + \tau) + y_{n-j}(t))$$ (B.14)

Then the equation for z can be obtained for any mode of perturbation, $k = 0, ...,$ $(N - 1)$, which requires the real part of z negative in order to maintain the linear stability. We obtain

$$\left(e^{\tau z} - 1\right)\left[e^{\tau z}z - \sum_{j=1}^{m}\gamma_j f_{2j} + \sum_{j=1}^{m}\gamma_j f_{3j}(e^{-ij\alpha_k} - e^{-i(j-1)\alpha_k})\right]$$

$$= \tau \sum_{j=1}^{m}\gamma_j f_{1j}(e^{-ij\alpha_k} - e^{-i(j-1)\alpha_k}) \qquad (B.15)$$

Similarly, the first and second order terms in the expression of z can be written respectively as

$$z_1 = \frac{\sum_{j=1}^{m}\gamma_j f_{1j}}{\sum_{j=1}^{m}\gamma_j f_{2j}} \qquad (B.16)$$

$$z_2 = \frac{(1 - \tau\sum_{j=1}^{m}\gamma_j f_{2j})z_1^2 - z_1\sum_{j=1}^{m}\gamma_j f_{3j} - \sum_{j=1}^{m}\gamma_j f_{1j}(j - 1/2)}{\sum_{j=1}^{m}\gamma_j f_{2j}} > 0 \qquad (B.17)$$

Thus the stable condition of homogenous flow is described as follows

$$\tau < \frac{1}{\sum_{j=1}^{m}\gamma_j f_{2j}} - \frac{\sum_{j=1}^{m}\gamma_j f_{1j}\sum_{j=1}^{m}\gamma_j f_{3j} + \sum_{j=1}^{m}\gamma_j f_{1j}(j - 1/2)\sum_{j=1}^{m}\gamma_j f_{2j}}{\left(\sum_{j=1}^{m}\gamma_j f_{1j}\right)^2} \qquad (B.18)$$

B.3 Model III

Model III further distinguishes the influences of speed and spacing, e.g. the speeds of preceding P vehicles and the spacings of preceding Q vehicles, i.e.

$$\dot{v}_n(t + \tau) = f(\bar{s}_n(t), v_n(t), \Delta\bar{v}_n(t)) \qquad (B.19)$$

where the follower will consider the average net vehicle-vehicle gap of P preceding vehicles as $\bar{s}_n = \sum_{p=1}^{P}\varphi_p s_{n-p+1}$, where the weighting coefficients satisfy $\sum_{p-1}^{P}\varphi_p = 1$. And the follower will consider the average vehicle-vehicle velocity difference of Q preceding vehicles as $\Delta\bar{v}_n = \sum_{q=1}^{Q}\psi_q\Delta v_{n-q+1}$, where the weighting coefficients satisfy $\sum_{q=1}^{Q}\psi_q = \Psi_Q$, and the scaling coefficient Ψ_Q might not equal to 1 because s_{n-p+1} and Δv_{n-q+1} have different dimensions. P can be different from Q because they can be regarded as independent statistic variables.

$$\ddot{y}_n(t+\tau) = f_1 \sum_{p=1}^{P} \varphi_p(y_{n-p}(t) - y_{n-p+1}(t)) + f_2 \dot{y}_n(t)$$

$$+ f_3 \sum_{q=1}^{Q} \psi_q(\dot{y}_{n-q+1}(t) - \dot{y}_{n-q}(t)) \qquad \text{(B.20)}$$

where $f_1 = \partial f/\partial \bar{s}_n(\bar{s}, \bar{v}, 0) \geqslant 0$, $f_2 = \partial f/\partial \bar{v}_n(\bar{s}, \bar{v}, 0) \leqslant 0$ and $f_3 = \partial f/\partial \Delta \bar{v}_n(\bar{s}, \bar{v}, 0) \leqslant 0$.

The corresponding difference equation is

$$\dot{y}_n(t+2\tau) - \dot{y}_n(t+\tau)$$

$$= \tau f_1 \sum_{p=1}^{P} \varphi_p(y_{n-p}(t) - y_{n-p+1}(t)) + f_2(y_n(t+\tau) - y_n(t))$$

$$+ f_3 \sum_{q=1}^{Q} \psi_q(y_{n-q+1}(t+\tau) - y_{n-q+1}(t) - y_{n-q}(t+\tau) + y_{n-q}(t)) \qquad \text{(B.21)}$$

The algebraic equation for z can be obtained as

$$\left(e^{\tau z} - 1\right)\left[e^{\tau z}z - f_2 + f_3 \sum_{q=1}^{Q} \psi_q\left(e^{-iq\alpha_k} - e^{-i(q-1)\alpha_k}\right)\right]$$

$$= \tau f_1 \sum_{p=1}^{P} \varphi_p\left(e^{-ip\alpha_k} - e^{-i(p-1)\alpha_k}\right) \qquad \text{(B.22)}$$

Expanding the first and second order in the expression of expending z, we have

$$z_1 = \sum_{p=1}^{P} \varphi_p f_1/f_2 \qquad \text{(B.23)}$$

$$z_2 = \left[(1 - f_2\tau)z_1^2 - f_3 z_1 \sum_{q=1}^{Q} \psi_q - f_1 \sum_{p=1}^{P} \varphi_p (p - 1/2)\right]/f_2 > 0 \qquad \text{(B.24)}$$

Thus, the stable condition of homogenous flow is described as follows

$$\tau < \frac{1}{f_2} - \frac{\sum_{q=1}^{Q} \psi_q f_3 + f_2 \sum_{p=1}^{P} \varphi_p (p - 1/2)}{f_1} \qquad \text{(B.25)}$$

References

Abul-Magd AY (2006) Modelling gap-size distribution of parked cars using random-matrix theory. Phys A 368(2):536–540

Abul-Magd AY (2007) Modeling highway-traffic headway distributions using superstatistics. Phys Rev E 76(5):057101

van Aerde M, Rakha H (2002) Integration release 2.30 for windows: User guide. Tech. rep., Virginia Tech., Transportation Institute.

Ahn S, Cassidy MJ, Laval J (2004) Verification of a simplified car-following theory. Transp Res Part B 38(5):431–440

Ahn S, Laval JA, Cassidy MJ (2010) Effects of merging and diverging on freeway traffic oscillations. Transp Res Rec 2188:1–8

Aw A, Rascle M (2000) Resurrection of second order models of traffic flow. SIAM J Appl Math 60(3):916–938

Aycin MF, Benekohal RF (1999) Comparison of car-following models for simulation. Transp Res Rec 1678:116–127

Bando M, Hasebe K, Nakayama A, Shibata A, Sugiyama Y (1995) Dynamical model of traffic congestion and numerical simulation. Phys Rev E 51(2):1035–1042

Banks JH (1989) Freeway speed-flow-concentration relationships: more evidence and interpretations. Transp Res Rec 1225:53–60

Banks JH (1991a) Two-capacity phenomenon at freeway bottlenecks: a basis for ramp metering. Transp Res Rec 1320:83–90

Banks JH (1991b) The two-capacity phenomenon: some theoretical issues. Transp Res Rec 1320:234–241

Banks JH (2002) Review of empirical research on congested freeway flow. Transp Res Rec 1802:225–232

Banks JH (2003) Average time gaps in congested freeway flow. Transp Res Part A 37(6):539–554

Banks JH (2006) New approach to bottleneck capacity analysis: final report. California PATH Research, California (Report pp UCB-ITS-PRR-2006-13)

Bassan S, Polus A, Faghri A (2006) Time-dependent analysis of density fluctuations and breakdown thresholds on congested freeways. Transp Res Rec 1965:40–47

Beaulieu NC, Rajwani F (2004) Highly accurate simple closed-form approximations to lognormal sum distributions and densities. IEEE Commun Lett 8(12):709–711

Beaulieu NC, Xie Q (2004) An optimal lognormal approximation to lognormal sum distributions. IEEE Trans Veh Technol 53(2):479–489

© Tsinghua University Press, Beijing and Springer-Verlag Berlin Heidelberg 2015   177
X. (M.) Chen et al., *Stochastic Evolutions of Dynamic Traffic Flow*,
DOI 10.1007/978-3-662-44572-3

Ben-Akiva M, Koutsopoulos HN, Antoniou C, Balakrishna R (2010) Traffic simulation with dyna-
    mit. In: Barcelo J (ed) Fundamentals of Traffic Simulation. Springer, New York
Ben Slimane S (2001) Bounds on the distribution of a sum of independent lognormal random
    variables. IEEE Trans Commun 49(6):975–978
Bertini RL, Leal MT (2005) Empirical study of traffic features at a freeway lane drop. J Transp Eng
    ASCE 131(6):397–407
Bierley RL (1963) Investigation of an intervehicle spacing display. Highway Res Rec 25:58–75
Boel R, Mihaylova L (2006) A compositional stochastic model for real time freeway traffic simu-
    lation. Transp Res Part B 40(4):319–334
Boer ER (1999) Car following from the driver's perspective. Transp Res Part F 2(4):201–206
Brackstone M, McDonald M (1999) Car-following: a historical review. Transp Res Part F 2(4):181–
    196
Brackstone M, McDonald M (2007) Driver headway: how close is too close on a motorway?
    Ergonomics 50(8):1183–1195
Brackstone M, Waterson B, McDonald M (2009) Determinants of following headway in congested
    traffic. Transp Res Part F 12(2):131–142
Branston D (1976) Models of single lane time headway distributions. Transp Sci 10(2):125–148
Brilon W, Geistefeldt J, Zurlinden H (2001) Implementing the concept of reliability for highway
    capacity analysis. Transp Res Rec 2027:1–8
Brilon W, Geistefeldt J, Regler M (2005) Reliability of freeway traffic flow: a stochastic concept of
    capacity. Proceedings of the 16th International Symposium of Transportation and Traffic Theory.
    College Park, Maryland, pp 125–144
Broughton KLM, Switzer F, Scott D (2007) Car following decisions under three visibility conditions
    and two speeds tested with a driving simulator. Accid Anal Prev 39(1):106–116
Buckley DJ (1968) A semi-poisson model of traffic flow. Transp Sci 2(2):107–133
Burghout W, Koutsopoulos H, Andréasson I (2006) Hybrid mesoscopic-microscopic traffic simu-
    lation. Transp Res Rec 1934:218–225
Cassidy MJ, Bertini RL (1999) Some traffic features at freeway bottlenecks. Transp Res Part B
    33(1):25–42
Cassidy MJ, Windover JR (1998) Driver memory: motorist selection and retention of individualized
    headways in highway traffic. Transp Res Part A 32(2):129–137
Cassidy MJ, Anani SB, Haigwood JM (2002) Study of freeway traffic near an off-ramp. Transp Res
    Part A 36(6):563–572
Chandler RE, Herman R, Montroll EW (1958) Traffic dynamics: studies in car following. Oper Res
    6:165–184
Chen A, Zhou Z (2010) The α-reliable mean-excess traffic equilibrium model with stochastic travel
    times. Transp Res Part B 44(4):493–513
Chen X, Li L, Jiang R, Yang XM (2010a) On the intrinsic concordance between the wide scattering
    feature of synchronized flow and the empirical spacing distributions. Chin Phys Lett 27(7):074501
Chen X, Li L, Zhang Y (2010b) A markov model for headway/spacing distribution of road traffic.
    IEEE Trans Intell Transp Syst 11(4):773–785
Chen X, Shi QX, Li L (2010c) Location specific cell transmission model for freeway traffic. Tsinghua
    Science and Technology in Press, Beijing
Chiabaut N, Buisson C, Leclercq L (2009) Fundamental diagram estimation through passing rate
    measurements in congestion. IEEE Trans Intell Transp Syst 10(2):355–359
Chiabaut N, Leclercq L, Buisson C (2010) From heterogeneous drivers to macroscopic patterns in
    congestion. Transp Res Part B 44(2):299–308
Choudhury CF (2007) Modeling driving decisions with latent plans. Ph.D. thesis, Massachusetts
    Institute of Technology, Cambridge, MA, USA
Chow AHF, Lu XY, Qiu TZ (2009) Empirical analysis of traffic breakdown probability distribution
    with respect to speed and occupancy. California PATH Working Paper pp UCB-ITS-PWP-2009-5
Chowdhury D, Santen L, Schadschneider A (2000) Statistical physics of vehicular traffic and some
    related systems. Phys Rep 329:199–329

Chowdhury D, Schadschneider A, Nishinari K (2010) Stochastic transport in complex systems: from molecules to vehicles. Elsevier, Oxford

Coifman B, Kim S (2011) Extended bottlenecks, the fundamental relationship, and capacity drop on freeways. Transp Res Part A 45(9):980–991

Coifman B, Dhoorjaty S, Lee ZH (2003) Estimating median velocity instead of mean velocity at single loop detectors. Transp Res Part C 11(3–4):211–222

Cowan RJ (1975) Useful headway models. Transp Res 9(6):371–375

Daganzo CF (1981) Estimation of gap acceptance parameters within and across the population from direct roadside observation. Transp Res Part B 15(1):1–15

Daganzo CF (1994) The cell transmission model: a dynamic representation of highway traffic consistent with the hydrodynamic theory. Transp Res Part B 28(4):269–287

Daganzo CF (1995a) The cell transmission model, part ii: network traffic. Transp Res Part B 29(2):79–93

Daganzo CF (1995b) A finite difference approximation of the kinematic wave model of traffic flow. Transp Res Part B 29(4):261–276

Daganzo CF (1995c) Requiem for second-order fluid approximations of traffic flow. Transp ResPart B 29(4):277–286

Daganzo CF (1999) The lagged cell transmission model. Proceeding of the 14th international symposium of transportation and traffic theory. Jerusalem, Israel, pp 147–171

Daganzo CF (2001) A simple traffic analysis procedure. Netw Spat Econ 1(1):77–101

Daganzo CF (2005) A variational formulation of kinematic waves: solution methods. Transp Res Part B 39(10):934–950

Daganzo CF, Cassidy MJ, Bertini RL (1999) Possible explanations of phase transitions in highway traffic. Transp Res Part A 33(5):365–379

Dailey DJ (1999) A statistical algorithm for estimating speed from single loop volume and occupancy measurements. Transp Res Part B 33(5):313–322

Dawson RF, Chimini LA (1968) The hyperlang probability distribution: a generalized traffic headway model. In: Highway Research Record, Highway Research Board, Washington, DC Characteristics of Traffic Flow(230). pp 58–66

Deco G, Scarano L, Soto-Faraco S (2007) Weber's law in decision making: integrating behavioral data in humans with a neurophysiological model. J Neurosci 27(42):11,192–11,200

Del Castillo JM (2001) Propagation of perturbations in dense traffic flow: a model and its implications. Transp Res Part B 35(4):367–389

Del Castillo JM, Benitez FG (1995a) On the functional form of the speed density relationship i: general theory. Transp Res Part B 29(5):373–389

Del Castillo JM, Benitez FG (1995b) On the functional form of the speed density relationship ii: empirical investigation. Transp Res Part B 29(5):391–406

van Der Hulst M (1999) Anticipation and the adaptive control of safety margins in driving. Ergonomics 42(2):336–345

Duret A, Buisson C, Chiabaut N (2008) Estimating individual speed-spacing relationship and assessing ability of newell's car-following model to reproduce trajectories. Transp Res Rec 2088:188–197

Duret A, Bouffier J, Buisson C (2010) Onset of congestion from low-speed merging maneuvers within free-flow traffic stream. Transp Res Rec 2188:96–107

Edie LC (1961) Car-following and steady-state theory for noncongested traffic. Oper Res 9(1):66–76

Elefteriadou L, Lertworawanich P (2003) Defining, measuring and estimating freeway capacity. In: Proceedings of the 82nd transportation research board annual meeting, Washington

Elefteriadou L, Roess R, McShane W (1995) Probabilistic nature of breakdown at freeway merge junctions. Transp Res Rec 1484:80–89

Elefteriadou L, Kondyli A, Brilon W, Hall FL, Persaud B, Washburn S (2011) Ramp metering enhancements for postponing freeway-flow breakdown. In: Proceedings of the 90th transportation research board annual meeting, Washington

Evans JL, Elefteriadou L, Gautam N (2001) Probability of breakdown at freeway merges using markov chains. Transp Res Part B 35(3):237–254

Evans L, Rothery R (1977) Perceptual thresholds in car following: a recent comparison. Transp Sci 11(1):60–72

Farrow K, Haag J, Borst A (2006) Nonlinear, binocular interactions underlying flow field selectivity of a motion-sensitive neuron. Nat Neurosci 9(10):1312–1320

Forbes TW (1963) Human factor consideration in traffic flow theory. Highway Res Rec 15:60–66

Frost BJ (2009) Lee's tau operator. Perception 38(6):852–853

Fuller R (2005) Towards a general theory of driver behaviour. Accid Anal Prev 37(3):461–472

Gawron C (1998) Simulation-based traffic assignment; computing user equilibria in large street networks. Ph.D. thesis, University of Cologne, Germany

Gazis DC, Herman R, Rothery RW (1961) Nonlinear follow-the-leader models of traffic flow. Oper Res 9:545–567

Geistefeldt J, Brilon W (2009) A comparative assessment of stochastic capacity estimation methods. In: Lam W, Wong S, Lo H (eds) Proceedings of the 18th international symposium of transportation and traffic theory. Springer, Hong Kong, China, pp 583–602

Gilchrist RS, Hall FL (1989) Three-dimensional relationships among traffic flow theory variables. Transp Res Rec 1225:109–115

Gipps PG (1981) A behavioural car-following model for computer simulation. Transp Res Part B 15(2):105–111

Gomes G, Horowitz R (2006) Optimal freeway ramp metering using the asymmetric cell transmission model. Transp Res Part C 14(4):244–262

Gomes G, Horowitz R, Kurzhanskiy AA, Varaiya P, Kwon J (2008) Behavior of the cell transmission model and effectiveness of ramp metering. Transp Res Part C 16(4):485–513

Greenberg JM (2001) Extensions and amplifications of a traffic model of aw and rascle. SIAM J Appl Math 62(3):729–745

Greenshields BD (1935) A study of highway capacity. Proc Highw Res Rec Wash 14:448–477

Habib-Mattar C, Polus A, Cohen MA (2009) Analysis of the breakdown process on congested freeways. Transp Res Rec 2124:58–66

Hall EL, Allen BL, Gunter MA (1986) Empirical analysis of freeway flowcdensity relationships. Transp Res Part A 20(3):197–210

Hall FL, Agyemang-Duah K (1991) Freeway capacity drop and the definition of capacity. Transp Res Rec 1320:91–98

Hamdar S, Mahmassani H (2008) From existing accident-free car-following models to colliding vehicles: exploration and assessment. Transp Res Rec 2088:45–56

Hancock PA (1999) Is car following the real question—are equations the answer? Transp Res Part F 2(4):197–199

Helbing D (2001) Traffic and related self-driven many-particle systems. Rev Mod Phys 73(4):1067–1141

Helbing D (2009) Derivation of non-local macroscopic traffic equations and consistent traffic pressures from microscopic car-following models. European Phys J B 69(6):1–10

Helbing D, Johansson AF (2009) On the controversy around daganzorequiem for and aw-rascles resurrection of second-order traffic flow models. European Phys J B 69:549–562

Helbing D, Tilch B (1998) Generalized force model of traffic dynamics. Phys Rev E 58(1):133–138

Helbing D, Hennecke A, Treiber M (1999) Phase diagram of traffic states in the presence of inhomogeneities. Phys Rev Lett 82(21):4360

Helbing D, Treiber M, Kesting A, Schünhof M (2009) Theoretical versus empirical classification and prediction of congested traffic states. European Phys J B 69:583–598

Herrera JC, Bayen AM (2010) Incorporation of lagrangian measurements in freeway traffic state estimation. Transp Res Part B 44(4):460–481

Hidas P (2005) Modelling vehicle interactions in microscopic simulation of merging and weaving. Transp Res Part C 13(1):37–62

Hoeffding W (1963) Probability inequalities for sums of bounded random variables. J Am Stat Assoc 58(301):13–30

Hoogendoorn S, Bovy P (2001) State-of-the-art of vehicular traffic flow modelling. Proc Inst Mech Eng Part I J Syst Control Eng 215(4):283–303

Hoogendoorn SP, Bovy PHL (1998) A new estimation technique for vehicle-type specific headway distribution. Transp Res Rec 1646:18–28

Hu S, Gao K, Wang B, Lu Y (2008) Cellular automaton model considering headway-distance effect. Chin Phys B 17(5):1863–1868

Huang P, Kong L, Liu M (2001) The study on the one-dimensional random traffic flow. Acta Phys Sin 50(1):30–36

Jabari SE, Liu HX (2012) A stochastic model of traffic flow: theoretical foundations. Transp Res Part B 46(1):156–174

Jayakrishnan R, Mahmassani HS, Hu TY (1994) An evaluation tool for advanced traffic information and management systems in urban networks. Transp Res Part C 2(3):129–147

Jiang R, Wu QS, Zhu ZJ (2001) Full velocity difference model for a car-following theory. Phys Rev E 64(1):017101

Jiang R, Wu QS, Zhu ZJ (2002) A new continuum model for traffic flow and numerical tests. Transp Res Part B 36(5):405–419

Jin XX, Zhang Y, Wang F, Li L, Yao D, Su Y, Wei Z (2009) Departure headways at signalized intersections: a log-normal distribution model approach. Transp Res Part C 17(3):318–327

Jost D, Nagel K (2003) Probabilistic traffic flow breakdown in stochastic car following models. Traffic and Granular Flow '03 pp 87–103

Kallenberg O (1997) Foundations of modern probability. Springer-Verlag, New York

Kalos MH, Whitlock PA (2008) Monte carlo methods, 2nd edn. Wiley, New York

van Kampen NG (2007) Stochastic processes in physics and chemistry, 3rd edn. Elsevier, Amsterdam

Kaplan EL, Meier P (1958) Nonparametric estimation from incomplete observations. J Am Stat Assoc 53:457–481

Kerner BS (2001) Complexity of synchronized flow and related problems for basic assumptions of traffic flow theories. Netw Spat Econ 1(1):35–76

Kerner BS (2004) The physics of traffic: empirical freeway pattern features, engineering applications, and theory. Springer-Verlog, New York

Kerner BS (2009) Introduction to modern traffic flow theory and control: the long road to three-phase traffic theory. Springer-Verlag, Berlin

Kerner BS, Klenov SL (2006) Probabilistic breakdown phenomenon at on-ramp bottlenecks in three-phase traffic theory: congestion nucleation in spatially non-homogeneous traffic. Phys A 364:473–492

Kerner BS, Rehborn H (1996a) Experimental features and characteristics of traffic jams. Phys Rev E 53(2):R1297

Kerner BS, Rehborn H (1996b) Experimental properties of complexity in traffic flow. Phys Rev E 53(5):R4275

Kesting A (2008) Microscopic modeling of human and automated driving: Towards traffic-adaptive cruise control. Ph.D. thesis, Technische Universität Dresden, Germany

Kesting A, Treiber M (2008) Calibrating car-following models by using trajectory data methodological study. Transp Res Rec 2088:148–156

Kesting A, Treiber M, Helbing D (2007) General lane-changing model mobil for car-following models. Transp Res Rec 1999:86–94

Kesting A, Treiber M, Helbing D (2010) Enhanced intelligent driver model to access the impact of driving strategies on traffic capacity. Philos Trans Roy Soc 368(1928):4585–4605

Kim T, Zhang HM (2008) A stochastic wave propagation model. Transp Res Part B 42(7–8):619–634

Knospe W, Santen L, Schadschneider A, Schreckenberg M (2002) Single-vehicle data of highway traffic: microscopic description of traffic phases. Phys Rev E 65(056):133

Kondyli A, Elefteriadou L (2011) Modeling driver behavior at freeway-ramp merges. Transp Res Rec 2249:29–37

Kondyli A, Soria I, Duret A, Elefteriadou L (2011) Sensitivity analysis of corsim with respect to the process of freeway flow breakdown at bottleneck locations. Simul Model Pract Theory 22:197–206

Krbalek M (2007) Equilibrium distributions in a thermodynamical traffic gas. J Phys A 40(22):5813–5821

Krbalek M, Helbing D (2004) Determination of interaction potentials in freeway traffic from steady-state statistics. Phys A 333:370–378

Krbalek M, Šeba P (2009) Spectral rigidity of vehicular streams (random matrix theory approach). J Phys A 42(345):001

Krbalek M, Šeba P, Wagner P (2001) Headways in traffic flow: remarks from a physical perspective. Phys Rev E 64(6):066119

Kühne R, Lüdtke A (2013) Traffic breakdowns and freeway capacity as extreme value statistics. Transp Res Part C 27:159–168

Kühne R, Mahnke R, Lubashevsky I, Kaupužs J (2002) Probabilistic description of traffic breakdowns. Phys Rev E 65(066):125

Kühne RD (1987) Freeway speed distribution and acceleration noise: calculations from a stochastic continuum theory and comparison with measurements. In: Gartner NH, Wilson NHM (eds) Proceedings of the 10th international symposium on transportation and traffic theory. Elsevier, New York, pp 119–137

Lam CLJ, Le-Ngoc T (2006) Estimation of typical sum of lognormal random variables using log shifted gamma approximation. IEEE Commun Lett 10(4):234–235

Laval JA, Leclercq L (2010a) Continuum approximation for congestion dynamics along freeway corridors. Transp Sci 44(1):87–97

Laval JA, Leclercq L (2010b) A mechanism to describe the formation and propagation of stop-and-go waves in congested freeway traffic. Philoso Trans Roy Soc A 368(1928):4519–4541

Lebacque JP (1996) The godunov scheme and what it means for first order traffic flow models. In: Proceedings of the 13th international symposium on transportation and traffic theory, Pergamon-Elservier, New York

Leclercq L, Chiabaut N, Laval J, Buisson C (2007) Relaxation phenomenon after lane changing: experimental validation with NGSIM data set. Transp Res Rec 1999:79–85

Leclercq L, Laval JA, Chiabaut N (2011) Capacity drops at merges: an endogenous model. Procedia Soc Behav Sci 7:12–26

Lee DN (1976) A theory of visual control of braking based on information about time-to-collision. Perception 5(4):437–459

Lee DN (2009) General tau theory: evolution to date. Perception 38(6):837–850

Lee JW (2004) Reversible random sequential adsorption on a one-dimensional lattice. Phys A 331(3):531–537

Lenz H, Wagner CK, Sollacher R (1999) Multi-anticipative car-following model. European Phys J B 7(2):331–335

Leonard DR, Power P, Taylor NB (1989) Contram: structure of the model. Technical Report, Transportation Research Laboratory (TRL), Crowthorn

Li B (2009) On the recursive estimation of vehicular speed using data from a single inductance loop detector: a bayesian approach. Transp Res Part B 43(4):391–402

Li B (2010) Bayesian inference for vehicle speed and vehicle length using dual-loop detector data. Transp Res Part B 44(1):108–119

Li J, Zhang HM (2011) Fundamental diagram of traffic flow: new identification scheme and further evidence from empirical data. Transp Res Rec 2260:50–59

Li J, Chen QY, Wang H, Ni D (2012) Analysis of LWR model with fundamental diagram subject to uncertainties. Transportmetrica 8(6):387–405

Li L, Wang FY (2007) Advanced motion control and sensing for intelligent vehicles. Springer-Verlag, New York

Li L, Wang F, Jiang R, Hu JM, Ji Y (2010a) A new car-following model yielding log-normal type headways distributions. Chin Phys B 19(2):020513

Li MZF (2008) A generic characterization of equilibrium speed-flow curves. Transp Sci 42(2):220–235

Li X, Kuang H, Song T, Dai S, Li Z (2008) New insights into traffic dynamics: a weighted probabilistic cellular automaton model. Chin Phys B 17(7):2366–2372

Li XP, Peng F, Ouyang Y (2010b) Measurement and estimation of traffic oscillation properties. Transp Res Part B 44(1):1–14

Lighthill MJ, Whitham GB (1955) On kinematic waves. ii. a theory of traffic flow on long crowded roads. Proc Roy Soc Lond Ser A Math Phys Sci 229(1178):317–345

Limpert E, Stahel WA, Abbt M (2001) Lognormal distributions across the sciences: keys and clues. Bioscience 51(5):341C352

Lin Y, Song H (2006) Dynachina: a real-time traffic estimation and prediction system. IEEE Persvasive Comput 5(4):65–72

Lipshtat A (2009) Effect of desired speed variability on highway traffic flow. Phys Rev E 79(066):110

Lo HK, Szeto WY (2002) A cell-based variational inequality formulation of the dynamic user optimal assignment problem. Transp Res Part B 36(5):421–443

Lo HK, Chang E, Chan YC (2001) Dynamic network traffic control. Transp Res Part A 35(8):721–744

Lorenz M, Elefteriadou L (2001) Defining freeway capacity as a function of breakdown probability. Transp Res Rec 1776:43–51

Lu L, Su Y, Yun T, Yao D, Li L (2010) A comparison of phase transitions produced by paramics, transmodeler and vissim. IEEE Intell Transp Syst Mag 2(3):19–24

Lu XY, Skabardonis A (2007) Freeway traffic shockwave analysis: exploring the NGSIM trajectory data. In: Proceeding of the 86th transportation research board annual meeting, Washington

Luttinen RT (1996) Statistical properties of vehicle time headway. Ph.D. thesis, Helsinki University of Technology, Finland

Ma X, Ingmar A (2006) Estimation of driver reaction time from car-following data: application in evaluation of general motor-type model. Transp Res Rec 1965:130–141

Maerivoet S, De Moor B (2005) Cellular automata models of road traffic. Phys Rep 419(1):1–64

Mahmassani H, Sheffi Y (1981) Using gap sequences to estimate gap acceptance functions. Transp Res Part B 15(3):143–148

Mahnke R, Kaupužs J (2001) Probabilistic description of traffic flow. Netw Spat Econ 1(1):103–136

Mahnke R, Kühne R (2007) Probabilistic description of traffic breakdown. Traffic and Granular Flow'05. pp 527–536

Mahnke R, Pieret N (1997) Stochastic master-equation approach to aggregation in freeway traffic. Phys Rev E 56(3):2666–2671

Mahnke R, Kaupužs J, Lubashevsky I (2005) Probabilistic description of traffic flow. Phys Rep 408(1–2):1–130

Mehmood A, Saccomanno F, Hellinga B (2003) Application of system dynamics in car-following models. J Transp Eng ASCE 129(6):625–634

Mehta NB, Wu J, Molisch AF, Zhang J (2007) Approximating a sum of random variables with a lognormal. IEEE Trans Wireless Commun 6(7):2690–2699

Michael PG, Leeming FC, Dwyer WO (2000) Headway on urban street: observational data and an intervention to decrease tailgating. Transp Res Part F 3(3):54–64

Michaels RM (1963) Perceptual factors in car following. In: Proceedings of the 2nd international symposium on the theory of road traffic flow. OECD, Paris, pp 44–59

Michaels RM, Solomon D (1962) Effect of speed change information on spacing between vehicles. Highw Res Board Bull 330:26–39

Muñoz L, Sun X, Horowitz R, Alvarez L (2003) Traffic density estimation with the cell transmission model. Am Control Conf. Denver, Colorado, pp 3750–3755

Muñoz L, Sun X, Horowitz R, Alvarez L (2006) Piecewise-linearized cell transmission model and parameter calibration methodology. Transp Res Rec 1965:183–191

Munjal PK, Pipes LA (1971) Propagation of on-ramp density waves on uniform unidirectional multilane freeways. Transp Sci 5(4):390–402

Nadarajah S (2008) A review of results on sums of random variables. Acta Applicandae Math 103(2):131–140

Nagel K, Schreckenberg M (1992) A cellular automaton model for freeway traffic. J Phys I 2(12):2221–2229

Nelsen RB (2006) An introduction to copulas. Springer Verlag, New York

Neubert L, Santen L, Schadschneider A, Schreckenberg M (1999) Single-vehicle data of highway traffic: a statistical analysis. Phys Rev E 60(6):6480–6490

Newell GF (1961) Nonlinear effects in the dynamics of car following. Oper Res 9(2):209–229

Newell GF (1965) Instability in dense highway traffic: A review. In: Almond J (ed) Proceedings of the 2nd international symposium of transportation and traffic theory. Organization for Economic Cooperation and Developmen, London, UK, pp 73–83

Newell GF (2002) A simplified car-following theory: a lower order model. Transp Res Part B 36(3):195–205

Ngoduy D (2011) Multiclass first-order traffic model using stochastic fundamental diagrams. Transportmetrica 7(2):111–125

NGSIM (2006) Next generation simulation. http://www.ngsim.fhwa.dot.gov/

Nishinari K, Treiber M, Helbing D (2003) Interpreting the wide scattering of synchronized traffic data by time gap statistics. Phys Rev E 68(6):067101

Ossen S (2008) Longitudinal driving nehavior: Theory and empirics. Ph.D. thesis, Delft University of Technology, Netherlands

Ossen S, Hoogendoorn S, Gorte B (2006) Interdriver differences in car-following: a vehicle trajectory-based study. Transp Res Rec 1965:121–129

Ossen SJ, Hoogendoorn SP (2005) Car-following behavior analysis from microscopic trajectory data. Transp Res Rec 1934:13–21

Ossen SJ, Hoogendoorn SP, Gorte BG (2007) Inter-driver differences in car-following: a vehicle trajectory based study. Transp Res Rec 1965:121–129

Ovuworie GC, Darzentas J, Mcdowell MRC (1980) Free movers, followers and others: a reconsideration of headway distributions. Traffic Eng Control 21(8):425–428

Ozbay K, Ozguven EE (2007) A comparative methodology for estimating the capacity of a freeway section. In: Proceedings of the 10th IEEE international conference on intelligent transportation systems, Seattle, WA, USA

Panwai S, Dia H (2005) Comparative evaluation of microscopic car-following behavior. IEEE Trans Intell Transp Syst 6(3):314–325

Payne HJ (1971) Models of freeway traffic and control. In: Bekey GA (ed) Math Models Public Syst. Simulation Council, La Jolla, pp 51–61

PeMS (2005) California performance measurement system. http://pems.eecs.berkeley.edu/

PeMS (2011) California performance measurement system. http://www.pems.dot.ca.gov

Persaud B, Yagar S, Brownlee R (1998) Exploration of the breakdown phenomenon in freeway traffic. Transp Res Rec 1634:64–69

Persaud B, Yagar S, Tsui D, Look H (2001) Breakdown-related capacity for freeway with ramp metering. Transp Res Rec 1748:110–115

Petr S (2008) Markov chain of distances between parked cars. J Phys A 41(12):122003

Petrov VV (1975) Sums of independent random variables. Springer, New York

Phillips WF (1979) A kinetic model for traffic flow with continuum implications. Transp Plann Technol 5(3):131–138

Pipes LA (1953) An operational analysis of traffic dynamics. J Appl Phys 24(3):274–281

Polus A, Pollatschek M (2002) Stochastic nature of freeway capacity and its estimation. Can J Civ Eng 29(6):842–852

Prigogine I, Herman R (1971) Kinetic theory of vehicular traffic. American Elsevier, New York

Punzo V, Borzacchiello MT, Ciuffo B (2011) On the assessment of vehicle trajectory data accuracy and application to the next generation simulation (NGSIM) program data. Transp Res Part C 19(6):1243–1262

Ranney TA (1999) Psychological factors that influence car-following and car-following model development. Transp Res Part F 2(4):213–219

Rawal S, Rodgers GJ (2005) Modelling the gap size distribution of parked cars. Phys A 346(3):621–630

Renyi A (1963) On a one-dimensional problem concerning random space filling. Sel Trans Math Stat Probab 4:203–218

Rice J, van Zwet E (2004) A simple and effective method for predicting travel times on freeways. IEEE Trans Intell Transp Syst 5(3):200–207

Richards PI (1956) Shockwaves on the highway. Oper Res 4:42–51

Romeo M, Da Costa V, Bardou F (2003) Broad distribution effects in sums of lognormal random variables. European Phys J B 32(4):513–525

Schönhof M, Helbing D (2007) Empirical features of congested traffic states and their implications for traffic modeling. Transp Sci 41(2):135–166

Schrater PR, Knill DC, Simoncelli EP (2001) Perceiving visual expansion without optic flow. Nature 410:816–819

Schuhl A (1955) The probability theory applied to distribution of vehicles on two-lane highways. Poisson and Traffic, Eno Foundation, Saugatuck pp 59–75

Shawky M, Nakamura H (2007) Characteristics of breakdown phenomenon in merging sections of urban expressways in japan. Transp Res Rec 2012:11–19

Shladover SE, Lu XY, Cody D, Nowakowski C, Qiu ZT, Chow A, OConnell J, Nienhuis J, Su D (2010) Effects of cooperative adaptive cruise control on traffic flow: testing drivers' choices of following distances. Simulation modelling practice and theory. California PATH, Institute of Transportation Studies, University of California, Berkeley, pp UCB-ITS-PRR-2010-2

Skog I, Handel P (2009) In-car positioning and navigation technologies: a survey. IEEE Trans Intell Transp Syst 10(1):4–21

Smilowitz KR, Daganzo CF (2002) Reproducible features of congested highway traffic. Math Comput Model 35(5–6):509–515

Snelder M (2009) A comparison between dynameq and indy. Tech. rep., CIRRELT-2009-48

Son B, Kim T, Kim HJ, Lee S (2004) Probabilistic model of traffic breakdown with random propagation of disturbance for its application. Lect Notes Comput Sci 3215:45–51

Soriguera F, Robusté F (2011) Estimation of traffic stream space mean speed from time aggregations of double loop detector data. Transp Res Part C 19(1):115–129

Sugiyama Y, Fukui M, Kikuchi M, Hasebe K, Nakayama A, Nishinari K, Tadaki S, Yukawa S (2008) Traffic jams without bottlenecks: Experimental evidence for the physical mechanism of the formation of a jam. New J Phys 10(033):001

Sultan B, Brackstone M, McDonald M (2004) Drivers' use of deceleration and acceleration information in car-following process. Transp Res Rec 1883:31–39

Sumalee A, Zhong RX, Pan TL, Szeto WY (2011) Stochastic cell transmission model (SCTM): a stochastic dynamic traffic model for traffic state surveillance and assignment. Transp Res Part B 45(3):507–533

Sun H, Frost BJ (1998) Computation of different optical variables of looming objects in pigeon nucleus rotundus neurons. Nat Neurosci 1(4):296–303

Szeto W (2008) Enhanced lagged cell-transmission model for dynamic traffic assignment. Transp Res Rec 2085:76–85

Taieb-Maimon M, Shinar D (2001) Minimum and comfortable driving headways: reality versus perception. Hum Factors 43(1):159–172

Talbot J, Tarjus G, van Tassel PR, Viot P (2000) From car parking to protein adsorption: An overview of sequential adsorption processes. Colloids Surf A 165(1):287–324

Taylor DH (1964) Drivers' galvanic skin response and the risk of accident. Ergonomics 7(4):439–451

Thiemann C, Treiber M, Kesting A (2008) Estimating acceleration and lane-changing dynamics from next generation simulation trajectory data. Transp Res Rec 2088:90–101

Toledo T, Zohar D (2007) Modeling duration of lane changes. Transp Res Rec 1999:71–78

Tolle JE (1976) Vehicular headway distributions: testing and results. Transp Res Rec 567:56–64

Transportation Research Board of the National Academies (2000) Highway Capacity Manual

Transportation Research Board of the National Academies (2010) Highway Capacity Manual 2010, Volume 1: Concepts, Chapter 1, 2 and 4; Volume 2: Uninterrupted flow

Treiber M, Helbing D (2003) Memory effects in microscopic traffic models and wide scattering in flow-density data. Phys Rev E 68(4):046119

Treiber M, Kesting A (2011) Evidence of convective instability in congested traffic flow: a systematic empirical and theoretical investigation. Transp Res Part B 45(9):1362–1377

Treiber M, Hennecke A, Helbing D (2000) Congested traffic states in empirical observations and microscopic simulations. Phys Rev E 62(2):1805–1824

Treiber M, Kesting A, Helbing D (2006a) Delays, inaccuracies and anticipation in microscopic traffic models. Phys A 360(1):71–88

Treiber M, Kesting A, Helbing D (2006b) Understanding widely scattered traffic flows, the capacity drop, and platoons as effects of variance-driven time gaps. Phys Rev E 74(1):016123

Treiber M, Kesting A, Helbing D (2010) Three-phase traffic theory and two-phase models with a fundamental diagram in the light of empirical stylized facts. Transp Res Part B 44(8–9):983–1000

Wagner P (2011) A time-discrete harmonic oscillator model of human car-following. European Phys J B 84(4):713–718

Wagner P, Nagel K (2008) Comparing traffic flow models with different number of phases. European Phys J B 63:315–320

Wang B, Adams TM, Jin W, Meng Q (2010a) The process of information propagation in a traffic stream with a general vehicle headway: a revisit. Transp Res Part C 18(3):367–375

Wang C, Coifman B (2008) The effect of lane-change maneuvers on a simplified car-following theory. IEEE Transa Intell Transp Syst 9(3):523–535

Wang F, Li L, Hu J, Ji Y, Ma R, Jiang R (2009a) A markov process inspired ca model of highway traffic. Int J Mod Phys C 20(1):117–131

Wang H, Wang W, Chen X, Chen J, Li J (2007) Experimental features and characteristics of speed dispersion in urban freeway traffic. Transp Res Rec 1999:150–160

Wang H, Rudy K, Li J, Ni D (2010b) Calculation of traffic flow breakdown probability to optimize link throughput. Appl Mathe Model 34(11):3376–3389

Wang Y, Papageorgiou M (2005) Real-time freeway traffic state estimation based on extended kalman filter: a general approach. Transp Res Part B 39(2):141–167

Wang Y, Papageorgiou M, Messmer A (2006) Renaissance—a unified macroscopic model-based approach to real-time freeway network traffic surveillance. Transp Res Part C 14(3):190–212

Wang Y, Papageorgiou M, Messmer A, Coppola P, Tzimitsi A, Nuzzolo A (2009b) An adaptive freeway traffic state estimator. Automatica 45(1):10–24

Wasielewski P (1974) An integral equation for the semi-poisson headway distribution model. Trans Sci 8(3):237–247

Whitham GB (1974) Linear and Nonlinear Waves. Wiley-interscience, New York

Wilde GJS (1982) The theory of risk homeostasis: implications for safety and health. Risk Anal 2(4):209–225

Wilson E (2008) Mechanisms for spatio-temporal pattern formation in highway traffic models. Philos Trans Roy Soc A 366(1872):2017–2032

Windover JR, Cassidy MJ (2001) Some observed details of freeway traffic evolution. Transp Res Part A 35(10):881–894

van Winsum W (1999) The human element in car following models. Transp Res Part F 2(4):207–211

Wu N (2002) A new approach for modeling of fundamental diagrams. Transp Res Part A 36(10):867–884

Wu X, Liu HX, Geroliminis N (2011) An empirical analysis on the arterial fundamental diagram. Transp Res Part A 45(1):255–266

Xue Y, Dai SQ (2003) Continuum traffic model with the consideration of two delay time scales. Phys Rev E 68(6):066123

Yan JJ, Lorv B, Li H, Sun HJ (2011) Visual processing of the impending collision of a looming object: time to collision revisited. J Vis 11(12):1–25

Yeo H (2008) Asymmetric microscopic driving behavior theory. Ph.D. thesis, Department of Civil and Environmental Engineering, University of California, Berkeley, USA

Yeo H, Skabardonis A (2009) Understanding stop-and-go traffic in view of asymmetric traffic theory. In: Lam WHK, Wong SC, Lo HK (eds) Proceedings of the 18th international symposium of transportation and traffic theory. Pergamon-Elservier, Hong Kong, China, pp 99–115

Young C, Rice J (2006) Estimating velocity fields on a freeway from low-resolution videos. IEEE Trans Intell Transp Syst 7(4):463–469

Zhang GH, Wang YH, Wei H, Chen Y (2007) Examining headway distribution models using urban freeway loop event data. Transp Res Rec 1999:141–149

Zhang HM (1998) A theory of nonequilibrium traffic flow. Transp Res Part B 32(7):485–498

Zhang HM (2002) A non-equilibrium traffic model devoid of gas-like behavior. Transp Res Part B 36(3):275–290

Zhang HM, Shen W (2009) Numerical investigation of stop-and-go traffic patterns upstream of freeway lane drop. Transp Res Rec 2124:3–17

Zhao L, Ding J (2007) Least squares approximations to lognormal sum distributions. IEEE Trans Veh Technol 56(2):991–997

Zhu HB, Ge HX, Dai SQ (2007) A new cellular automaton model for traffic flow with different probability for drivers. Int J Mod Phys C 18(5):773–782

Index

© Tsinghua University Press, Beijing and Springer-Verlag Berlin Heidelberg 2015
X. (M.) Chen et al., *Stochastic Evolutions of Dynamic Traffic Flow*,
DOI 10.1007/978-3-662-44572-3

189